Engineering Applications of Computational Methods

Volume 25

The book series Engineering Applications of Computational Methods addresses the numerous applications of mathematical theory and latest computational or numerical methods in various fields of engineering. It emphasizes the practical application of these methods, with possible aspects in programming. New and developing computational methods using big data, machine learning and AI are discussed in this book series, and could be applied to engineering fields, such as manufacturing, industrial engineering, control engineering, civil engineering, energy engineering and material engineering.

The book series Engineering Applications of Computational Methods aims to introduce important computational methods adopted in different engineering projects to researchers and engineers. The individual book volumes in the series are thematic. The goal of each volume is to give readers a comprehensive overview of how the computational methods in a certain engineering area can be used. As a collection, the series provides valuable resources to a wide audience in academia, the engineering research community, industry and anyone else who are looking to expand their knowledge of computational methods.

This book series is indexed in both the **Scopus** and **Compendex** databases.

Jin Yi · Jun Zheng · Xinyu Li · Liang Gao

Variable-Fidelity Surrogate

Experiment Design, Modeling, and Applications on Design Optimization

Springer

Jin Yi
Chongqing University
Chongqing, China

Jun Zheng
China University of Geosciences
Wuhan, Hubei, China

Xinyu Li
School of Mechanical Science
and Engineering
Huazhong University of Science
and Technology
Wuhan, Hubei, China

Liang Gao
School of Mechanical Science
and Engineering
Huazhong University of Science
and Technology
Wuhan, Hubei, China

ISSN 2662-3366 ISSN 2662-3374 (electronic)
Engineering Applications of Computational Methods
ISBN 978-981-95-5526-0 ISBN 978-981-95-5527-7 (eBook)
https://doi.org/10.1007/978-981-95-5527-7

This Springer imprint is published by the registered company Springer Nature Singapore Pte Ltd.
The registered company address is: 152 Beach Road, #21-01/04 Gateway East, Singapore 189721, Singapore

Preface

In recent years, numerical simulations have become essential tools in engineering design and optimization, enabling the analysis and replication of complex real-world systems. Advances in computational power and simulation software have facilitated the creation of more detailed models, thereby reducing the discrepancy between simulated and actual behaviors. However, high-fidelity simulations continue to demand significant computational resources, driven by increasing model complexity such as finer meshes and coupled multiphysics equations, making their direct use in iterative optimization processes often impractical.

In this context, surrogate models have attracted considerable interest as efficient approximations, built from sampled input-output data to provide rapid predictions. Variable-fidelity surrogate models, which integrate data from simulations of varying accuracy, offer particular advantages in optimization by effectively balancing predictive accuracy and computational cost. These methods have proven effective in reducing the number of required simulations and streamlining the design process.

This monograph provides a comprehensive overview of variable-fidelity surrogate models in engineering design and optimization. The text first addresses foundational concepts and construction techniques, including an introduction to surrogate modeling basics, experimental design methods for nested sampling, and advanced variable-fidelity modeling approaches such as scaling functions, physics-informed neural networks, and transfer learning-based surrogates. Subsequently, the focus shifts to their application in expensive optimization problems, encompassing surrogate-assisted metaheuristic algorithms for single-objective optimization and sequential strategies for multi-objective optimization. Practical examples from engineering fields, such as pressure vessel design, permanent magnet gears, and structural mechanics, illustrate the benefits of these approaches.

Emphasizing ensemble and variable-fidelity techniques, this book aims to address the trade-off between surrogate accuracy and computational efficiency. It serves as a valuable reference for graduate students, researchers, and engineers engaged in engineering design, optimization, and computational modeling, offering both theoretical foundations and practical guidance.

We extend our thanks to the following graduate students for their contributions: Songqing Xing, Wenjie Ding, Xuguang Liu, Hanhai Shou, and Xinze Li. Jin Yi, the first author, acknowledges financial support from the National Natural Science Foundation of China (Grant No. 52105244), the Entrepreneurship and Innovation Support Plan of Chongqing for Returned Overseas Scholars, China (Grant Nos. cx2023085 and cx2024027), and the Independent Research Project-Key Program from the State Key Laboratory of Mechanical Transmission for Advanced Equipment, China (Grant No. SKLMT-ZZKT-2024Z09). Jun Zheng, the second author, gratefully acknowledges funding from the National Natural Science Foundation of China (Grant No. 51505439).

Chongqing, China
June 2025

Jin Yi
Jun Zheng
Xinyu Li
Liang Gao

Contents

Acronyms

ANN	Artificial Neural Networks
CFD	Computational Fluid Dynamics
DNN	Deep Neural Network
DOE	Design of Experiments
ED	Experimental Design
EGO	Efficient Global Optimization
EOS	Ensemble of Surrogates
EUCB	Extended Upper Confidence Bound
FEM	Finite Element Modeling
HDMR	High-Dimensional Model Representation
HF	High-fidelity
HSA	Harmony Search Algorithm
LF	Low-fidelity
LHD	Latin Hypercube Design
LVGP	Latent Variable Gaussian Process
MARS	Multivariate Adaptive Regression Spline
MCS	Monte Carlo Sampling
MDO	Multidisciplinary Design Optimization
MFEI	Multi-fidelity Expected Improvement
OAD	Orthogonal Arrays Design
PEI	Proportional Expected Improvement
PINN	Physics-Informed Neural Network
RBF	Radial Basis Function
RSM	Response Surface Method
SBDO	Surrogate-Based Design Optimization
SVR	Support Vector Regression
TR-AMMO	Trust Region-Approximation Model Management Optimization
UD	Uniform Design
VFS	Variable-fidelity Surrogate
VFSM	Variable-fidelity Surrogate Modeling

Chapter 1
Introduction

1.1 Modern Product Design Optimization

In the design of complex engineering products, such as aircraft, ships, and geotechnical machinery, computational simulation technologies (e.g., finite element analysis (FEM), computational fluid dynamics (CFD)) have become indispensable tools. These simulation techniques replace expensive physical experiments, shorten development cycles, and provide essential data for design optimization. As computer performance advances, the discretization of finite elements in simulations has become more refined, significantly improving computational accuracy. This enhancement enables more precise and realistic simulations, leading to higher-fidelity computational models. However, this increased accuracy often leads to a dramatic rise in computational time required for these models [1]. Literature reports that Volvo's D5 diesel engine optimization required 40 hours per simulation run [2]. The multi-objective genetic algorithm involved approximately 20,000 runs, requiring about 91 years for optimization. Ford's automobile crash simulation took between 26 and 160 hours per run [3]. A two-variable design optimization requiring 50 simulations would take between 75 days and 11 months. This demonstrates that using such computationally intensive models directly for design optimization is often impractical due to time constraints. Computational models for complex engineering products are typically black-box models, where outputs are known, but the functional relationship remains unknown. This black-box nature of computational simulation models restricts the applicability of gradient-based optimization methods.

Surrogate-based design optimization (SBDO) methods, emerging in the 1980s, have gained significant attention. Surrogate models, also called metamodels, are simple numerical models constructed by fitting or interpolating input-output data. They are typically used to replace expensive simulation models. Since surrogate models capture the functional relationship between inputs and outputs, they can be directly employed for optimization. They replace costly simulations across the design space, enabling rapid estimation of responses at unknown design points. With

J. Yi et al., *Variable-Fidelity Surrogate*, Engineering Applications of Computational Methods 25, https://doi.org/10.1007/978-981-95-5527-7_1

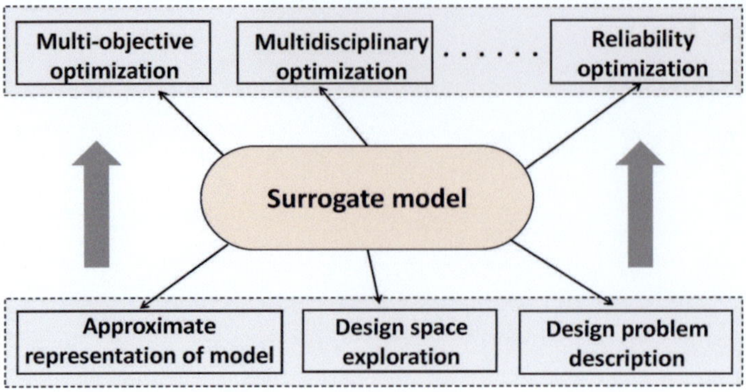

Fig. 1.1 Applications of surrogate models in design optimization problems

explicit input-output relationships, engineers can easily perform sensitivity analysis, parameter experimentation, and design space exploration. This also helps engineers better understand the design problem characteristics. Sensitivity analysis, parameter experimentation, and design space exploration reduce the number of design variables, eliminate ineffective constraints, and may simplify multiple objectives into a single one. Surrogate models address the challenges of high computational density and time costs in design optimization, aiding in multidisciplinary, robustness, reliability, and global optimization [4], as shown in Fig. 1.1.

Constructing surrogate models follows a bottom-up process, consisting of experimental design and modeling. It involves collecting input-output data from the original simulation model, then fitting or interpolating to establish the functional relationship and form the surrogate model. Surrogate models do not require knowledge of the black-box nature of the original model, making them applicable across fields like ecology [5], economics [6], physics [7], and engineering [8]. Commonly used surrogate models include Support Vector Regression (SVR) [9], Kriging models [10], Response Surface Method (RSM) [11], Artificial Neural Networks (ANN) [12], Multivariate Adaptive Regression Spline (MARS) [13], and Radial Basis Function (RBF) models [14]. These surrogate models are typically constructed by sampling the computationally expensive high-fidelity analysis model and are considered single-fidelity models.

Fidelity refers to how closely physical laws or assumptions in analysis models align with reality. For example, in finite element simulations, the mesh may be coarse or fine [15]. High-fidelity models contain more physical details, offering more accuracy but requiring greater computational effort. Low-fidelity models use simplified assumptions, reducing computational cost but potentially sacrificing accuracy. Constructing single-fidelity surrogate models typically requires a large number of samples (ranging from hundreds to thousands) from the high-fidelity analysis model to ensure accuracy, resulting in substantial simulation time costs.

To address this issue, researchers have introduced analysis models of varying fidelities and constructed surrogate models by integrating data from different fidelity levels. This approach, known as variable-fidelity surrogate models, was proposed around 2000 [16], and is also referred to as the hierarchical model. Similar to conventional single-fidelity surrogate models, variable-fidelity surrogate models sample the analysis model that needs to be approximated and establish a function relating the model inputs and outputs. These models are widely applicable in fields such as aerospace, geotechnical mechanics, multidisciplinary design optimization, and reliability-based design optimization.

What distinguishes variable-fidelity surrogate models from single-fidelity models is their ability to leverage both high- and low-fidelity models. These models utilize a large number of low-fidelity samples to reduce computational complexity and simulation time while relying on a few high-fidelity samples to improve accuracy.

The introduction of low-fidelity models significantly reduces the sampling and the number of runs required for high-fidelity analysis models, thereby greatly decreasing the time cost of model construction. This enables the creation of high-accuracy surrogate models using a small number of high-fidelity samples, which has garnered considerable attention from researchers [9, 17, 18]. Figure 1.2 illustrates the applications of variable-fidelity surrogate models, showing their relevance across multiple industries such as automation, chemical engineering, aerospace, electronics, and materials. These models can be integrated with techniques such as model-based design, optimization, sensitivity analysis, and CAx-based design.

In conclusion, the design and optimization of modern complex engineering products often involve high-fidelity simulation models with high computational density, leading to significant computational time costs, which pose challenges for design

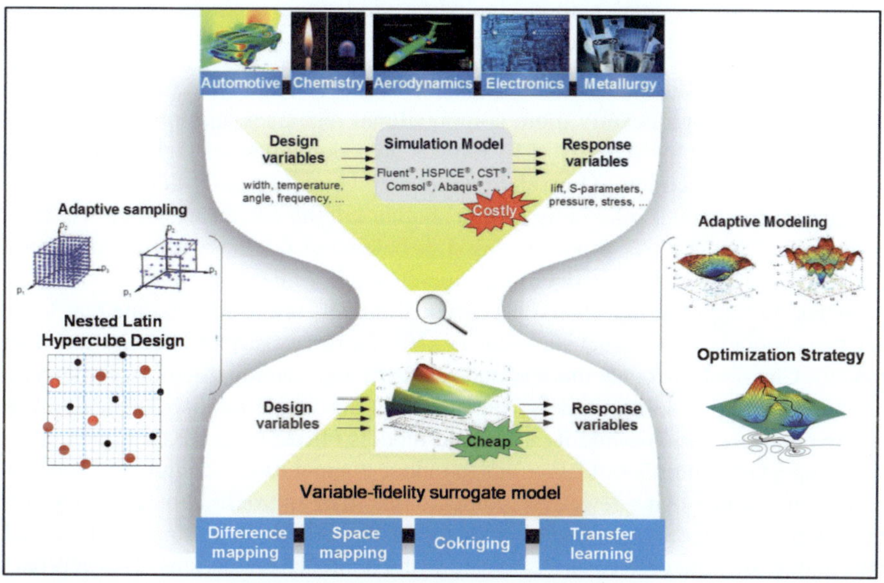

Fig. 1.2 Applications of variable-fidelity surrogate models

optimization. By using surrogate models to replace computationally expensive high-fidelity simulation models, the computational time costs in design optimization can be substantially reduced. As an emerging surrogate model, the variable-fidelity surrogate model shows great potential by integrating information from both high- and low-fidelity analysis models, allowing for the construction of low-cost, high-precision surrogate models. This has resulted in widespread attention and research in this area.

1.2 Variable-Fidelity Surrogate Models

In practical engineering scenarios, designers have access to both high-fidelity models, which accurately describe the problem, and lower-fidelity models, which provide approximate representations. For example, the mesh resolution in finite element models can vary. "Fidelity" refers to how accurately an analytical model represents the original problem, reflecting the alignment of the model's physical assumptions with real-world conditions. In engineering design optimization, models of varying fidelity typically arise from the following sources [19, 20]:

1. Different physical or mathematical descriptions, such as laminar and turbulent flow models or the Navier-Stokes equations for viscous Newtonian fluids versus the Euler equations for inviscid flows.
2. Variations in resolution, such as coarse or fine meshes in finite element analysis.
3. Discrepancies in accuracy, such as the use of simplified boundary conditions or the adoption of two-dimensional models to approximate three-dimensional ones.
4. Different experimental methods, including wind tunnel testing, CFD simulations, and engineering estimation techniques.

High-fidelity (HF) models offer more detailed information and greater accuracy, but they require significantly more computational resources and longer simulation times. In contrast, low-fidelity (LF) models are computationally cheaper but less accurate than HF models. Thus, the relationship between model fidelity and computational efficiency is inherently inverse: higher fidelity correlates with lower computational efficiency. As shown in Fig. 1.3, balancing fidelity with computational efficiency is a key challenge in engineering design optimization.

When multiple fidelity levels of analytical models are available, one can leverage these models to construct surrogate models, known as variable-fidelity surrogate models. Classical surrogate models, such as RSM and Kriging, focus on simplifying or sampling high-fidelity models, thus representing single-fidelity surrogate models. In contrast, variable-fidelity surrogate models utilize both HF and LF models, sampling a larger number of LF points to reduce computational complexity and simulation time while using fewer HF points to maintain approximation accuracy. The inclusion of LF models significantly reduces the frequency of invoking the HF model, thus minimizing the time costs associated with HF model evaluations. This approach addresses the challenge of building accurate surrogate models with a limited number of samples. Consequently, variable-fidelity surrogate-based design optimization has

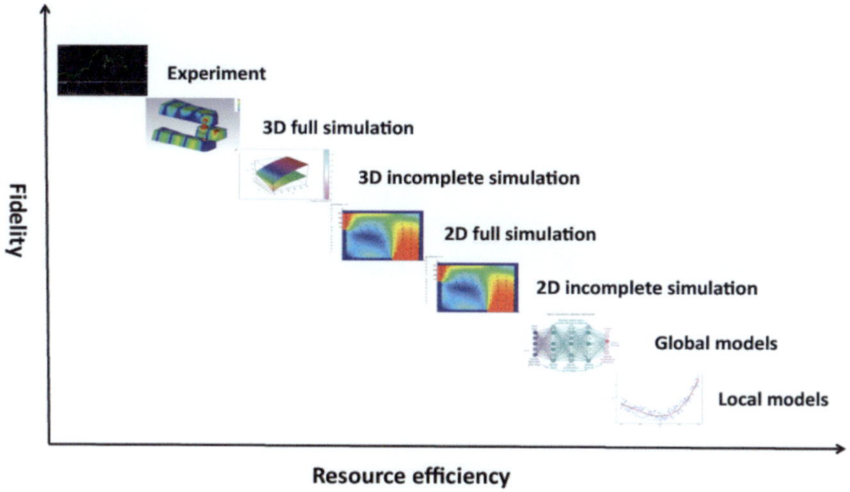

Fig. 1.3 Relationship between model fidelity and computational efficiency

emerged as a powerful method to reduce computational costs while ensuring reliable optimization results.

Figure 1.4 illustrates a comprehensive framework for the overall process. The research primarily focuses on two key areas: the construction of the mathematical model and the solution process. Specifically, this includes model construction, experimental design, optimization strategy, and solution algorithms.

(1) **Design of Experiments (DOE)**: The DOE is a method developed to efficiently organize experiments that yield reliable information with the fewest trials. In the context of experimental design, the data obtained from samples at various points in the design space provide diverse insights into the model. The goal of DOE is to strategically sample from positions in the design space that yield the most information with minimal sampling. Common DOE methods include uniform design [21], Taguchi design [22], orthogonal sampling [23], and Latin hypercube design [24]. In the context of variable-fidelity surrogate modeling, samples are drawn from simulations with different fidelity levels, resulting in a two-layer experimental design that includes both HF and LF samples. Traditional DOE methods, such as orthogonal sampling, which are designed for single-fidelity surrogate models, are not directly applicable to experimental design in variable-fidelity models.
(2) **Evaluation**: At each sample point in the experimental design, both HF and LF simulation models are run, producing corresponding HF and LF model outputs (i.e., target function responses) that are used to construct the variable-fidelity surrogate model.
(3) **Variable-Fidelity Surrogate Models**: These models use the HF and LF information obtained from the evaluation process to approximate the objective or con-

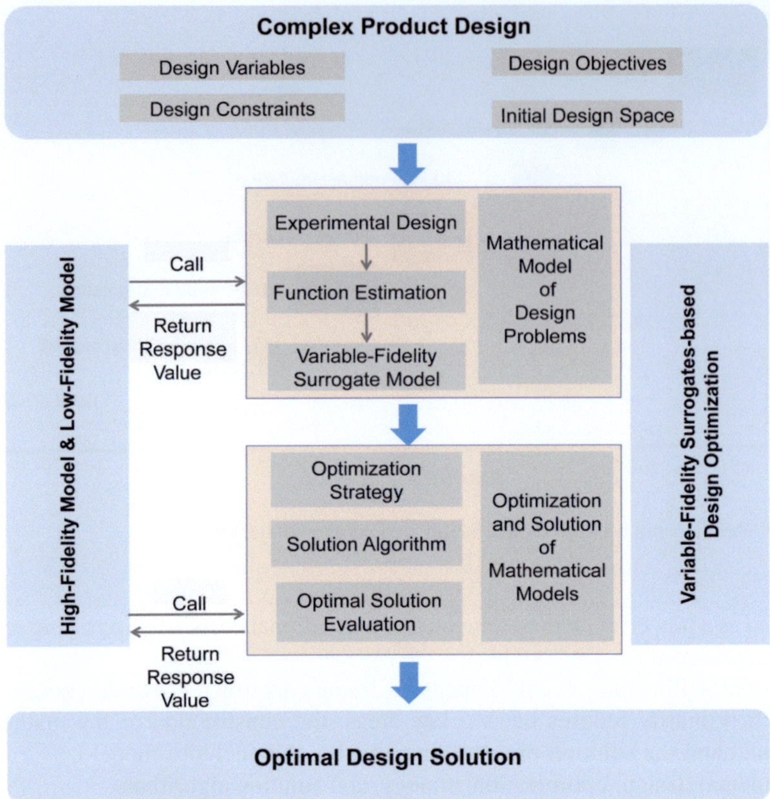

Fig. 1.4 Main research areas in variable-fidelity surrogate modeling

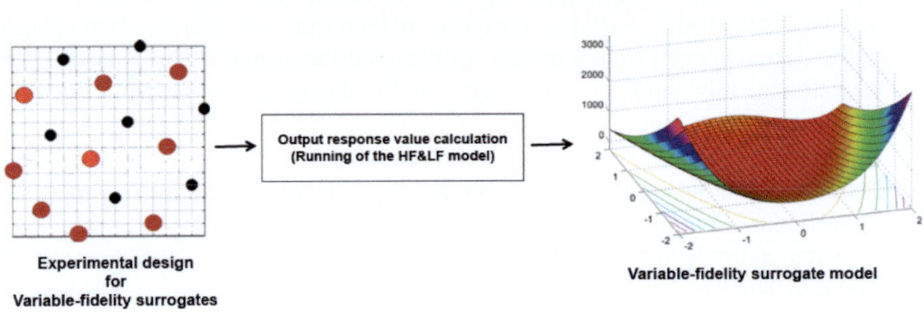

Fig. 1.5 Schematic of the construction process for variable-fidelity surrogate models

straint functions. Compared to single-fidelity surrogate models, variable-fidelity surrogate models make full use of the strengths of both HF and LF models. By utilizing a greater number of LF samples to reduce computational complexity and simulation time, and a smaller number of HF samples to maintain accuracy, variable-fidelity surrogate models significantly reduce model construction costs.

(4) **Optimization Strategy**: To address the trade-off between computational efficiency and optimization accuracy in complex engineering design problems, an iterative process based on variable-fidelity surrogate models is developed. This process integrates the advantages of both HF and LF models, forming an optimization framework to resolve the challenge of high computational demand in designing complex engineering products.

(5) **Optimization Algorithms**: Optimization algorithms are employed to solve the problem and obtain the optimal solution. Common optimization algorithms include deterministic methods and heuristic intelligent optimization algorithms.

(6) **Optimal Solution Evaluation**: The optimal solution derived from the optimization process is evaluated using the HF model to verify its accuracy.

The points outlined above summarize the key research areas in variable-fidelity surrogate model construction. The core focus is on developing variable-fidelity surrogate models and the associated optimization strategies.

Figure 1.5 presents a schematic of the process for constructing a variable-fidelity surrogate model. This process integrates the characteristics of different fidelity models, simplifying complex objective and constraint functions into more tractable forms. In constructing a variable-fidelity surrogate model, it is crucial to balance model accuracy with construction costs. Therefore, experimental designs tailored for variable-fidelity surrogates are of paramount importance.

Figure 1.6 illustrates the optimization strategy based on variable-fidelity surrogates. This strategy not only enhances optimization performance but also uses HF model information to assess the optimization results while updating and refining the LF surrogate model. During optimization, different fidelity models transition

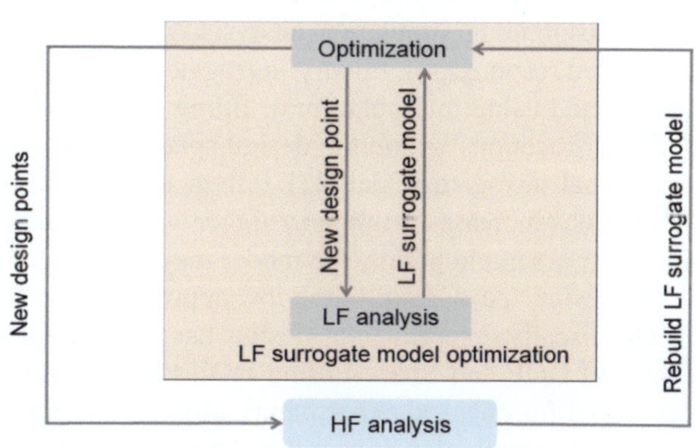

Fig. 1.6 Schematic of the optimization strategy based on variable-fidelity surrogates

between roles, from high to low fidelity, ensuring that the optimization process is more efficient, with higher computational accuracy and reduced costs.

Thus, the variable-fidelity surrogate-based optimization process consists of several key steps:

1. Experimental design methods suited for variable-fidelity surrogate models.
2. Construction of the variable-fidelity surrogate model.
3. Formulation of optimization strategies.

In-depth analysis and synthesis of these three research areas are critical to advancing the field of variable-fidelity surrogate modeling.

1.3 Literature Review on Variable-Fidelity Surrogate Modeling and Applications

This section reviews and discusses the key techniques in constructing variable-fidelity surrogate models, focusing on experimental design methods, surrogate modeling approaches, and optimization strategies for variable-fidelity approximations.

1.3.1 Review on Experimental Design for Variable-Fidelity Surrogates

Experimental Design
Experimental design is the process of determining the distribution of experimental sampling points within the design space. It is the first step in constructing surrogate models and marks the beginning of the entire surrogate-based design optimization process. To build a surrogate model, sampling points must first be determined through experimental design within the design space, forming the set of design points. Then, the corresponding simulation analysis is run at these design points to obtain the associated response set. Finally, surrogate models of the original analysis model are constructed using interpolation or fitting techniques. The number of design points and their locations within the design space significantly influence the performance of the final surrogate model [25]. If there are too many sample points, the number of simulations increases, leading to higher sampling time and cost. Conversely, if there are too few sample points, the model may not adequately reflect the characteristics of the design space, resulting in low approximation accuracy.

Historically, experimental design has been divided into two major categories: Classical DOE and Space-Filling DOE [26]. Classical DOE, which was primarily developed for physical experiments, aims to minimize random errors in the experiments. The distribution of sample points in classical experimental design is typically concentrated at the boundaries of the design space, with very few points at the center, which is why it is also referred to as edge-distributed experimental design. Classical experimental design methods include Full/Fractional Factorial Design [27], Central Composite Design (CCD) [28], D-Optimal Design (DO) [29, 30], and Plackett-Burman designs [31, 32]. With the development of computer technology, computer experiments have started to replace physical experiments in certain cases. Unlike physical experiments, computer experiments involve systematic errors rather than random errors [33]. A good experimental design should fill the entire design space, rather than focusing solely on the boundaries [34]. For deterministic computer experiments, classic designs such as CCD and DO are inefficient and may not be suitable. For these types of experiments, sample points should cover the entire design space, which is the essence of X (also known as modern experimental design) [35–37]. Common space-filling design methods include Monte Carlo Sampling (MCS) [38], Uniform Design (UD) [39], Hammersley Designs [40], Orthogonal Arrays Design (OAD) [39, 41], and Latin Hypercube Design (LHD) [24, 42].

Experimental Design for Variable-Fidelity Surrogates

In the process of constructing variable-fidelity surrogate models, sample information can be obtained using different types of fidelity analysis models. Although LF models can quickly provide sample information, the accuracy of the obtained data is limited. In contrast, HF models require significant human, material, and financial resources to obtain sample information, but they offer higher precision. During experimental design, both high- and low-fidelity models must be configured, and the correlation between the sample points from these two fidelities must be identified.

The key problem to be solved in the design of a variable-fidelity surrogate model is how to arrange high/low precision sample points in the design space to provide reasonable and effective data for the construction of a variable-fidelity surrogate model. Yi et al. [43] used harmony search and local enumeration methods to construct a nested experimental design with the maximum distance. Lin et al. [44] proposed a sequential sampling method for robust design optimization based on multi-fidelity modeling, which considers both the uncertainty of design variables and the uncertainty of interpolation in the sequence sampling process to realize the sequential sampling experiment design. Hu et al. [45] proposed an efficient multi-objective optimization method to improve the space-filling and orthogonality of LHD. Zheng [46] proposed a nested Latin Hypercube method based on translation propagation, which does not rely on intelligent algorithms and is easier to implement due to the replication of translation seeds. Theoretically, the experimental designs found by these methods may be optimal for certain criteria, but they are highly dependent on algorithm performance. As the number of experimental design points increases, analyzing all HF points from LF becomes impractical due to time constraints. In recent years, adaptive sampling strategies, which can increase sample points based on model information in a targeted manner, have been widely combined with variable-fidelity surrogate models to further improve model construction efficiency. Xiao et al. proposed an adaptive variable-fidelity surrogate model for uncertainty propagation analysis [47]. Wackers et al. [48] presented a novel initialization strategy with a limited training set for adaptive sampling. It allows the surrogate model to adapt more freely to the data and improves model accuracy. Zhang et al. [49] developed an expectation incremental method based on variable-fidelity surrogate information. Zheng [50] proposed an adaptive sampling strategy based on dynamically changing confidence boundaries. It can be observed that combining adaptive sampling with variable-fidelity surrogate models can utilize model information to guide the selection of the next sample points, which helps to improve model accuracy. However, the computational cost is higher than that of a one-time sampling approach, and the quality of the selected sample points heavily depends on the initial sample selection.

1.3.2 Review on Variable-Fidelity Surrogate Models

Surrogate Models

The concept of surrogate models was first introduced by researcher Kleijnen [51], who applied this concept to replace computational models in complex engineering product design processes, including CFD and FEM. In simple terms, surrogate models can be categorized into local surrogates and global surrogates. The former approximates the objective or constraint functions in the vicinity of a specific design point, with common methods such as Taylor series expansion [52]. The latter approximates functions across the entire design space and includes a variety of models, such as Response Surface Method (RSM) [11], Radial Basis Function (RBF) [14, 53], Artificial Neural Networks (ANN) [12], Support Vector Regression (SVR) [9], and Multivariate Adaptive Regression Spline (MARS) [13].

To achieve high prediction accuracy, large-scale sampling, i.e., experimental design, is required before constructing these surrogate models. No single surrogate model is universally superior to others. Each model has its own characteristics and is suitable for specific situations. Kriging is particularly accurate in approximating low-order and high-dimensional linear and nonlinear functions. It can flexibly select interpolation sample points or filter noisy data, but its construction process is complex and requires solving a global optimization problem using maximum likelihood estimation [37]. RSM models fit sample points using least squares to approximate corresponding functions. These models are simple to construct and operate, making them suitable for approximating low-order, low-dimensional linear functions, but they are less accurate for nonlinear functions [3]. RBF can be used for approximating high-order, high-dimensional linear and nonlinear functions, but its accuracy is lower than that of Kriging [3]. MARS, with its unique basis function generation method, can conveniently and accurately approximate high-dimensional functions, but its calculation time is slow and the precision is low [54]. SVR has a broad application range, suitable for high-order, medium-low-dimensional nonlinear problems, balancing robustness and approximation accuracy, and has potential for wide applications [55].

Variable-Fidelity Surrogate Models

Once different fidelity levels of physical phenomena are obtained through analysis models, these fidelity models can be utilized to construct a surrogate model. This is known as a variable-fidelity surrogate model. The main task of a variable-fidelity surrogate model is to precisely articulate the interplay between HF and LF models [56]. According to the type of assumption relation, the modeling method can be divided into three types: linear relationship assumption, nonlinear relationship assumption, and no prior assumption. The method based on the linear relationship assumption realizes data fusion through the linear bridge function between HF and LF, which is widely used in variable-fidelity surrogate modeling. In early studies, Zhu et al. [57] used the Kriging model as a multiplicative bridge function to correct the LF model. Zhou et al. [58] used SVR to construct an additive bridge function to optimize the process parameters. Yi et al. [59] adopted the comprehensive bridge function and established a variable-fidelity surrogate model based on the RBF model. Toal et al. [9] extended the optimization scenario of multi-output Kriging to mixed integers based on the traditional multiplicative bridge function, and verified its multi-output optimization potential. Lai et al. [60] proposed a variable-fidelity surrogate modeling method through local adaptive scaling factors based on the correlation of design variables.

Although the linear method is computationally efficient, its ability to characterize the complex nonlinear relationship between HF and LF is limited, especially in high-dimensional or strongly nonlinear scenarios. In order to overcome the limitations of linear methods, nonlinear methods have been developed rapidly in recent years.

Perdikaris et al. [61] proposed a probabilistic framework based on Gaussian process regression and nonlinear autoregression to learn the nonlinear and spatial correlation between HF and LF. Meng et al. [62] improved the universality of the model by adapting the degree of nonlinearity through the weighted fusion of linearity and nonlinearity. Zhang et al. [63] established a variable-fidelity surrogate model based on variable-fidelity deep neural networks. A low-fidelity model is first created through a deep neural network, and its output is then used as an input to the neural network to train the variable-fidelity deep neural networks along with high-fidelity data. Li et al. [64] proposed a nonlinear integrated double fidelity model to enhance the learning ability of nonlinear correlation. Zhang et al. [65] proposed a multi-level hierarchical kriging model and optimized the aerodynamic structure of a helicopter rotor, which successfully achieved drag reduction. Ma et al. [66] fit the high-frequency and low-frequency sample points, respectively, and then built a collaborative Kriging model for the optimization of EDM process parameters. Huang et al. [67] proposed a variable-fidelity surrogate model based on Co-Kriging and proposed a proportional expectation improvement method to improve the efficiency of global optimization. As the research continues, researchers begin to explore efficient surrogate modeling methods when there are more fidelity levels. Leng et al. [68] constructed three models with distinct fidelity levels, subsequently implementing hierarchical transfer learning across these levels. While conventional variable-fidelity surrogate modeling requires explicit hierarchical distinctions between models, practical applications frequently encounter scenarios where fidelity hierarchies are inseparable. To address the fusion of non-hierarchical LF data, Chen et al. [69] proposed a localized correlation-weighted fusion method, which significantly enhanced the utilization efficiency of multi-source data.

With the advancement of machine learning technologies, deep neural networks (DNNs) have demonstrated remarkable linear and nonlinear approximation capabilities. This has prompted researchers to employ deep learning for autonomously capturing complex relationships between high- and low-fidelity models. Conti et al. [70] proposed a data-driven variable-fidelity surrogate modeling framework based on long short-term memory networks in the context of parameterized time-varying systems. This framework demonstrated improved output prediction capabilities for unknown parameter values and enhanced temporal extrapolation performance through sequential learning mechanisms. Yi et al. [56] used deep transfer learning technology to propose a weighted training strategy of deep transfer and dynamic task to improve the prediction accuracy of high-dimensional problems. Chen et al. [71] proposed an innovative Latent Variable Gaussian Process (LVGP) model aimed at capturing correlations among variable-fidelity models without requiring any assumptions about hierarchical structures between fidelity levels. Zhou et al. [72] addressed the low computational efficiency of traditional Physics-Informed Neural Networks (PINNs) in high-dimensional spaces by introducing a Multi-Scale Physics-Constrained Neural Network (MSPCNN). This framework enhances fluid dynamics predictions through variable-fidelity autoencoders for integrated data assimilation.

1.3.3 Review on Optimization Strategies

Surrogate-Based Design Optimization (SBDO) Strategies

SBDO refers to the use of surrogate models in the design optimization process. There are primarily two SBDO strategies, as shown in Fig. 1.7. The first is the offline SBDO strategy, which involves experimental design, surrogate model construction, and optimization. If model validation is incorporated, cross-validation is often used. This approach typically requires a substantial number of sample points and performs optimization directly after model construction without further adjustments. It is simple, and suitable for optimizing various surrogate models [73–75]. Figure 1.7b illustrates an online SBDO strategy, which introduces a phase for sample adjustment and model reconstruction. Based on model validation or optimization data, this phase determines whether sample points need to be adjusted and the model reconstructed.

Current research on offline SBDO is mainly focused on practical applications. For instance, Lee et al. [76] proposed an experimental surrogate-based design optimization of wing geometry and structure for flapping wing micro air vehicles. An optimal wing was obtained via a surrogate-based global optimization reflecting the noise generated from the experiments.

In the case of online SBDO, much of the research revolves around space reduction techniques. These methods construct surrogate models with improved approximation accuracy within a reduced design space and then perform optimization to identify

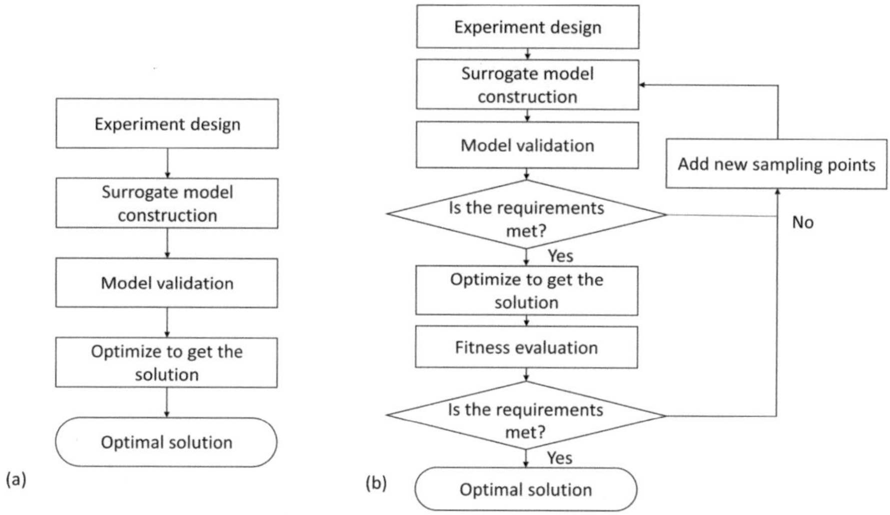

Fig. 1.7 Surrogate-based design optimization strategies **a** Offline SBDO, **b** Online SBDO

the optimal solution [77]. Alternatively, sequential sampling techniques are used to iteratively enhance the surrogate model's performance, such as improving optimal values or reducing errors, to ultimately obtain the optimal solution [78, 79].

Design Optimization Based on Variable-Fidelity Surrogates

Early studies on design optimization with variable-fidelity surrogates were primarily based on heuristic approaches. As early as 2000, Alexandrov proposed a surrogate model management strategy, with the Approximation Model Framework (AMF) being widely applied in aerospace fields, such as wing optimization and aircraft shape design [80, 81]. However, the convergence of AMF has not been theoretically validated. Based on the concept of trust regions, Alexandrov introduced the Trust Region-Approximation Model Management Optimization (TR-AMMO) strategy [82]. In TR-AMMO, optimization is performed around the current design point, and the approximation quality is evaluated using the trust region. When first-order consistency is satisfied, TR-AMMO can be proven to converge. Since the introduction of TR-AMMO, it has garnered considerable attention from the research community. Burton improved the gradient acquisition method, streamlining trust region control, thus optimizing TR-AMMO [83]. Hino applied TR-AMMO to optimize the blanking problem in sheet metal forming, resulting in a significant reduction in computational costs [84]. Rodriguez utilized TR-AMMO in studies on both finite and infinite molecular dynamics models, successfully developing a composite ceramic material with enhanced fracture toughness [85]. Silva applied TR-AMMO to enhance the performance of a gas turbine engine controller [86].

From the above review, it can be concluded that TR-AMMO effectively updates and manages variable-fidelity surrogate models using trust region-based rates, and its convergence has been proven. This method has been widely used. However, it is based on local approximations, making it unsuitable for multimodal problems. Therefore, global variable-fidelity surrogate-based optimization strategies have been developed. However, further research is needed to control the model update process effectively, reduce the use of HF models, and improve optimization accuracy.

The idea of efficient global optimization is to balance the relationship between global exploration and local search by adding sample points in sequence. Sequential optimization enhances the accuracy of variable-fidelity surrogate models and improves optimization efficiency through iterative updates of sample distributions, with its core lying in acquisition function design and sequential sampling strategies [87]. Xiong et al. [88] proposed a sequential nested Latin hypercube design. Ruan et al. [89] proposed a modified Probability of Improvement (PoI) criterion that integrates model correlation and cost ratios for sample selection. Shu et al. [90] developed an Expected Further Improvement (EFI) criterion based on multi-level Kriging, balancing information value with computational costs. Cheng et al. [91] established objective and constraint-oriented infill sampling functions, introducing a variable-fidelity constrained lower confidence bound criterion to enhance sampling efficiency. Huang et al. [67] proposed a Proportional Expected Improvement (PEI) method that optimizes global search through adaptive scaling factors and KL divergence. Han et al. [92] integrated multi-level hierarchical Kriging with Multi-Fidelity Expected Improvement (MFEI), effectively reducing computational costs in robust airfoil optimization. Lin et al. [44] developed an Extended Upper Confidence Bound (EUCB) method to address design variable and interpolation uncertainties. Pellegrini et al. [93] established an optimization framework based on variable-fidelity surrogate models and active learning, employing hierarchical error correction and stochastic radial basis functions to dynamically select training points and fidelity levels. Hai et al. [94] optimized high-fidelity sample selection through geometric transformations and active learning strategies. For expensive multi-objective optimization, Shu et al. [90] developed a framework integrating hierarchical Kriging-based multi-fidelity surrogate models with Bayesian optimization for expensive multi-objective optimization problems, designing a novel acquisition function balancing information value and sampling costs. Xue et al. [95] proposed a Co-Kriging-based variable-fidelity expected volume improvement method, which was successfully applied to aerospace structural multi-objective design.

References

1. Venkataraman S, Haftka R (2002) Structural optimization: what has Moore's law done for us? In: 43rd AIAA/ASME/ASCE/AHS/ASC structures, structural dynamics, and materials conference, p 1342
2. Jakobsson S, Patriksson M, Rudholm J et al (2010) A method for simulation based optimization using radial basis functions. Optim Eng 11:501–532
3. Jin R, Chen W, Simpson TW (2001) Comparative studies of metamodelling techniques under multiple modelling criteria. Struct Multidiscip Optim 23:1–13

4. Alizadeh R, Allen JK, Mistree F (2020) Managing computational complexity using surrogate models: a critical review. Res Eng Design 31(3):275–298
5. Esmaeilion F, Soltani M (2024) Technical and economic assessments of a novel multigeneration system based on desalination and liquid air energy storage. Desalination 580:117497
6. Hoseinzadeh S, Assareh E, Riaz A et al (2023) Ocean thermal energy conversion (OTEC) system driven with solar-wind energy and thermoelectric based on thermo-economic analysis using multi-objective optimization technique. Energy Rep 10:2982–3000
7. Bhosekar A, Ierapetritou M (2018) Advances in surrogate based modeling, feasibility analysis, and optimization: a review. Comput & Chem Eng 108:250–267
8. Yan C, Yin Z, Shen X et al (2020) Surrogate-based optimization with improved support vector regression for non-circular vent hole on aero-engine turbine disk. Aerosp Sci Technol 96:105332
9. Toal DJJ (2023) Applications of multi-fidelity multi-output Kriging to engineering design optimization. Struct Multidiscip Optim 66(6):125
10. Kleijnen JPC (2009) Kriging metamodeling in simulation: a review. Eur J Oper Res 192(3):707–716
11. Zheng H, Jiang B, Zheng Y et al (2023) Experimental study on forced ventilation and dust-control in a heading face based on response surface method. Process Saf Environ Prot 175:753–763
12. Villarrubia G, De Paz JF, Chamoso P et al (2018) Artificial neural networks used in optimization problems. Neurocomputing 272:10–16
13. Mushtaq K, Zou R, Waris A et al (2023) Multivariate wind power curve modeling using multivariate adaptive regression splines and regression trees. PLoS ONE 18(8):e0290316
14. Ji W, Fang L, Dong G et al (2023) State of health estimation of lithium-ion battery with improved radial basis function neural network. Energy 262:125380
15. Fernández-Godino MG (2023) Review of multi-fidelity models. Advances in Computational Science and Engineering 1. LLNL-JRNL-847790
16. Alexandrov N, Lewis R, Gumbert C et al (2000) Optimization with variable-fidelity models applied to wing design. In: 38th aerospace sciences meeting and exhibit, p 841
17. Feng D-C, Chen S-Z, Taciroglu E (2024) Deep learning-enhanced efficient seismic analysis of structures with multi-fidelity modeling strategies. Comput Methods Appl Mech Eng 421:116775
18. Liu X, Zhao W, Wan D (2022) Multi-fidelity Co-Kriging surrogate model for ship hull form optimization. Ocean Eng 243:110239
19. Do B, Zhang R (2023) Multifidelity Bayesian optimization: a review. AIAA J 1–37
20. Zheng J, Qiu H, Feng H (2012) The variable fidelity optimization for simulation-based design: a review. In: Proceedings of the 2012 IEEE 16th international conference on computer supported cooperative work in design (CSCWD). IEEE, pp 289–294
21. Fang K-T, Lin DKJ, Winker P et al (2000) Uniform design: theory and application. Technometrics 42(3):237–248
22. Huang X, Li F, Li Y et al (2023) Optimization of melting performance of a heat storage tank under rotation conditions: based on taguchi design and response surface method. Energy 271:127100
23. Burrage K, Burrage P, Donovan D et al (2015) Populations of models, experimental designs and coverage of parameter space by Latin hypercube and orthogonal sampling. Proc. Comput. Sci. 51:1762–1771
24. Viana FAC (2016) A tutorial on Latin hypercube design of experiments. Qual Reliab Eng Int 32(5):1975–1985
25. Zhou X, Gao Y, Jiang T et al (2023) An online approach for robust parameter design with incremental Gaussian process. Qual Eng 35(3):430–443
26. Myers RH, Montgomery DC, Anderson-Cook CM (2016) Response surface methodology: process and product optimization using designed experiments. Wiley
27. Kenett RS, Zacks S, Gedeck P (2023) Classical design and analysis of experiments. Industrial statistics: a computer-based approach with python. Springer, pp 141–224

28. Szpisják-Gulyás N, Al-Tayawi AN, Horváth ZH et al (2023) Methods for experimental design, central composite design and the Box-Behnken design, to optimise operational parameters: A review. Acta Aliment 52(4):521–537
29. Fernandes P, de Aguiar B, Bourguignon MSK et al (1995) D-optimal designs. Chemom Intell Lab Syst 30(2):199–210
30. Jones B, Allen-Moyer K, Goos P (2021) A-optimal versus D-optimal design of screening experiments. J Qual Technol 53(4):369–382
31. Deng L-Y, Tang B (1999) Generalized resolution and minimum aberration criteria for Plackett-Burman and other nonregular factorial designs. Statistica Sinica 1071–1082
32. Gregor M, Grznar P, Mozol S et al (2024) Plackett-Burman design. Acta Simulatio 10:1
33. Garud SS, Karimi IA, Kraft M (2017) Design of computer experiments: a review. Comput & Chem Eng 106:71–95
34. Susannah SJ, Schiller B, Welch WJ (1989) Designs for computer experiments. Technometrics 31(1):41–47
35. Lu L, Anderson-Cook CM, Martin M et al (2022) Practical choices for space-filling designs. Qual Reliab Eng Int 38(3):1165–1188
36. Giunta A, Wojtkiewicz S, Eldred M (2003) Overview of modern design of experiments methods for computational simulations. In: 41st aerospace sciences meeting and exhibit, p 649
37. Wang GG, Shan S (2006) Review of metamodeling techniques in support of engineering design optimization. In: International design engineering technical conferences and computers and information in engineering conference, vol 4255, pp 415–426
38. Hammersley J (2013) Monte carlo methods. Springer Science & Business Media
39. Wan K, Ma C (2016) Orthogonal and uniform experimental design. Science Press
40. Donmez M, Karamangil MI (2024) Artificial neural networks-based multi-objective optimization of immersion cooling battery thermal management system using Hammersley sampling method. Case Stud Therm Eng 64:105509
41. Panda A (2023) Orthogonal array design based multi-objective CBO and SOS algorithms for band reduction in hyperspectral image analysis. Multimedia Tools Appl 82(23):35301–35327
42. Yang J, Zeping W, Wang Z et al (2023) Enhanced anisotropic radius basis function metamodel based on recursive evolution Latin hypercube design and fast K-fold cross-validation. Struct Multidiscip Optim 66(7):169
43. Yi J, Li X, Xiao M et al (2017) Construction of nested maximin designs based on successive local enumeration and modified novel global harmony search algorithm. Eng Optim 49(1):161–180
44. Lin Q, Zhou Q, Jiexiang H et al (2022) A sequential sampling approach for multi-fidelity surrogate modeling-based robust design optimization. J Mech Des 144(11):111703
45. Xiaoru H, Ma P, Li S et al (2025) The optimization of space-filling and orthogonality for Latin hypercube design using a local search-based NSGA-II. J Stat Theory Practice 19(1):1–26
46. Zheng J (2018) An output mapping variable fidelity metamodeling approach based on nested Latin hypercube design for complex engineering design optimization. Appl Intell 48(10):3591–3611
47. Xiao W, Shen Y, Zhao J et al (2025) An adaptive multi-fidelity surrogate model for uncertainty propagation analysis. Appl Sci 15(6):3359
48. Wackers J, Pellegrini R, Serani A et al (2023) Efficient initialization for multi-fidelity surrogate-based optimization. J Ocean Eng Marine Energy 9(2):291–307
49. Zhang Y, Han Z-H, Zhang K-S (2018) Variable-fidelity expected improvement method for efficient global optimization of expensive functions. Struct Multidiscip Optim 58:1431–1451
50. Zheng J, Li Z, Gao L et al (2016) A parameterized lower confidence bounding scheme for adaptive metamodel-based design optimization. Eng Comput 33(7):2165–2184
51. Thode HC (1988) Statistical tools for simulation practitioners
52. Hilali Y, Sitli Y, Elmhaia O et al (2025) An improved high-order Kriging mesh-free approach for nonlinear thermal buckling of porous FG beams. Acta Mechanica 1–21
53. Chen G, Zhang K, Xue X et al (2022) A radial basis function surrogate model assisted evolutionary algorithm for high-dimensional expensive optimization problems. Appl Soft Comput 116:108353

54. Menéndez-García LA, García-Nieto PJ, García-Gonzalo E et al (2024) Time series analysis for COMEX platinum spot price forecasting using SVM, MARS, MLP, VARMA and ARIMA models: A case study. Resour Policy 95:105148
55. Ahmed HU, Mostafa RR, Mohammed A et al (2023) Support vector regression (SVR) and grey wolf optimization (GWO) to predict the compressive strength of GGBFS-based geopolymer concrete. Neural Comput Appl 35(3):2909–2926
56. You L, Xing S, Yi J et al (2025) A harmonic domain regressor with dynamic task weighting strategy for multi-fidelity surrogate modeling in engineering design. Adv Eng Inform 64:102999
57. Zhu J, Wang Y-J, Collette M (2014) A multi-objective variable-fidelity optimization method for genetic algorithms. Eng Optim 46(4):521–542
58. Zhou Q, Yang Y, Jiang P et al (2017) A multi-fidelity information fusion metamodeling assisted laser beam welding process parameter optimization approach. Adv Eng Softw 110:85–97
59. Yi J, Gao L, Li X et al (2019) An on-line variable-fidelity surrogate-assisted harmony search algorithm with multi-level screening strategy for expensive engineering design optimization. Knowl-Based Syst 170:1–19
60. Lai X, Pang Y, Liu F et al (2024) A multi-fidelity surrogate model based on design variable correlations. Adv Eng Inform 59:102248
61. Perdikaris P, Raissi M, Damianou A et al (2017) Nonlinear information fusion algorithms for data-efficient multi-fidelity modelling. Proc R Soc A Math Phys Eng Sci 473(2198):20160751
62. Meng X, Karniadakis GE (2020) A composite neural network that learns from multi-fidelity data: application to function approximation and inverse PDE problems. J Comput Phys 401:109020
63. Zhang X, Xie F, Ji T et al (2021) Multi-fidelity deep neural network surrogate model for aerodynamic shape optimization. Comput Methods Appl Mech Eng 373:113485
64. Li K, Li Q, Lv L et al (2023) A nonlinearity integrated bi-fidelity surrogate model based on nonlinear mapping. Struct Multidiscip Optim 66(9):196
65. Zhang Y, Han Z, Song W (2024) Multi-fidelity expected improvement based on multi-level hierarchical kriging model for efficient aerodynamic design optimization. Eng Optim 56(12):2408–2430
66. Ma J, Yin C, Li X et al (2022) Optimization of EDM process parameters based on variable-fidelity surrogate model. Int J Adv Manuf Technol 122(3):2031–2041
67. Huang H, Liu Z, Zheng H et al (2023) A proportional expected improvement criterion-based multi-fidelity sequential optimization method. Struct Multidiscip Optim 66(2):30
68. Leng J-X, Feng Y, Huang W et al (2024) Variable-fidelity surrogate model based on transfer learning and its application in multidisciplinary design optimization of aircraft. Phys Fluids 36:1
69. Chen H, Ouyang L, Liu L et al (2024) A multi-fidelity surrogate modeling method in the presence of non-hierarchical low-fidelity data. Aerosp Sci Technol 146:108928
70. Conti P, Guo M, Manzoni A et al (2023) Multi-fidelity surrogate modeling using long short-term memory networks. Comput Methods Appl Mech Eng 404:115811
71. Chen Y-P, Wang L, Comlek Y et al (2024) A latent variable approach for non-hierarchical multi-fidelity adaptive sampling. Comput Methods Appl Mech Eng 421:116773
72. Zhou H, Cheng S, Arcucci R (2024) Multi-fidelity physics constrained neural networks for dynamical systems. Comput Methods Appl Mech Eng 420:116758
73. Vardhan H, Hyde D, Timalsina U et al (2024) Sample-efficient and surrogate-based design optimization of underwater vehicle hulls. Ocean Eng 311:118777
74. Belen A, Günes F, Palandoken M et al (2022) 3D EM data driven surrogate based design optimization of traveling wave antennas for beam scanning in X-band: an application example. Wireless Netw 28(4):1827–1834
75. Fathima Sana VK, Nazeeh KM, Dilip DM et al (2022) Reliability-based design optimization of shallow foundation on cohesionless soil based on surrogate-based numerical modeling. Int J Geomech 22(2):04021283
76. Lee J, Yoon S-H, Kim C (2022) Experimental surrogate-based design optimization of wing geometry and structure for flapping wing micro air vehicles. Aerosp Sci Technol 123:107451

77. Zheng L, Yang Y, Fu G et al (2022) A surrogate-based optimization method with dynamic adaptation for high-dimensional mixed-integer problems. Swarm Evol Comput 72:101099
78. Park JW, Lee I (2023) A new framework for efficient sequential sampling-based RBDO using space mapping. J Mech Des 145(3):031702
79. Hu W, Zhao F, Deng X et al (2024) A new sequential sampling method for surrogate modeling based on a hybrid metric. J Mech Des 146:6
80. Xia L, Zheng-Hong G (2006) Application of variable-fidelity models to aerodynamic optimization. Appl Math Mech 27(8):1089–1095
81. Thokala P, Martins JRRA (2007) Variable-complexity optimization applied to airfoil design. Eng Optim 39(3):271–286
82. Alexandrov NM, Lewis RM, Gumbert CR et al (2001) Approximation and model management in aerodynamic optimization with variable-fidelity models. J Aircr 38(6):1093–1101
83. Burton SA, Hajela P (2003) A variable-complexity approach to second-order reliability-based optimization. Struct Multidiscip Optim 25:237–250
84. Hino R, Yoshida F, Toropov VV (2006) Optimum blank design for sheet metal forming based on the interaction of high-and low-fidelity FE models. Arch Appl Mech 75:679–691
85. Robinson TD, Eldred MS, Willcox KE et al (2008) Surrogate-based optimization using multifidelity models with variable parameterization and corrected space mapping. AIAA J 46(11):2814–2822
86. Silva VVR, Fleming PJ, Sugimoto J et al (2008) Multiobjective optimization using variable complexity modelling for control system design. Appl Soft Comput 8(1):392–401
87. Li K, Li F (2024) Multi-fidelity methods for optimization: a survey. arXiv:2402.09638
88. Xiong S, Qian PZG, Wu CFJ (2013) Sequential design and analysis of high-accuracy and low-accuracy computer codes. Technometrics 55(1):37–46
89. Ruan X, Jiang P, Zhou Q et al (2020) Variable-fidelity probability of improvement method for efficient global optimization of expensive black-box problems. Struct Multidiscip Optim 62:3021–3052
90. Shu L, Jiang P, Wang Y (2021) A multi-fidelity Bayesian optimization approach based on the expected further improvement. Struct Multidiscip Optim 63(4):1709–1719
91. Cheng J, Lin Q, Yi J (2022) An enhanced variable-fidelity optimization approach for constrained optimization problems and its parallelization. Struct Multidiscip Optim 65(7):188
92. Zhang Y, Han Z-H, Song W (2024) An efficient robust aerodynamic design optimization method based on a multi-level hierarchical Kriging model and multi-fidelity expected improvement. Aerospace Sci Technol 152:109401
93. Pellegrini R, Wackers J, Broglia R et al (2023) A multi-fidelity active learning method for global design optimization problems with noisy evaluations. Eng Comput 39(5):3183–3206
94. Hai C, Qian W, Wang W et al (2024) Active learning-assisted multi-fidelity surrogate modeling based on geometric transformation. Comput Methods Appl Mech Eng 426:116990
95. Xue T, Chen L, Hu J et al (2023) A variable-fidelity multi-objective optimization method for aerospace structural design optimization. Eng Optim 55(7):1133–1148

Chapter 2
Key Technologies for Surrogate Modeling

2.1 Surrogate Models

The fundamental idea of surrogate models (also called approximation models) is to construct a mathematical model that can simulate and approximate the input-output relationship of the original analysis model using a certain set of sample data and information. The relationship between input variables and the outputs of both the surrogate model and the original analysis model can be expressed as

$$y(x) = \hat{y}(x) + \varepsilon, \tag{2.1}$$

where x represents the input design variables, $y(x)$ is the actual response value, $\hat{y}(x)$ is the surrogate model's response value, and ε is the error between the approximated and actual values. The surrogate model $\hat{y}(x)$ is a relatively simple explicit functional formula or a mapping relationship with black-box attributes, which can effectively reduce computational costs and improve efficiency in complex engineering product optimization.

As discussed in the previous chapter, the key workflow of surrogate model technology consists of three stages: experiment of design, surrogate model construction, and optimization solving. Real-world complex engineering product optimization problems are subject to diverse and complex influencing factors, making it difficult to find a universal surrogate model framework applicable to all such problems. Therefore, appropriate surrogate model techniques should be judiciously selected for specific complex design optimization problems.

J. Yi et al., *Variable-Fidelity Surrogate*, Engineering Applications of Computational Methods 25, https://doi.org/10.1007/978-981-95-5527-7_2

2.1.1 Common Surrogate Models

2.1.1.1 Response Surface Methodology (RSM)

Response Surface Methodology (also known as polynomial regression), due to its simple principle and ease of implementation, is widely used. Its primary form is

$$\hat{y}(x) = \beta_0 + \sum_i \beta_i x_i + \sum_i \sum_{j>i} \beta_{ij} x_i x_j + \sum_i \beta_{ii} x_i^2 + \sum_i \sum_{j>i} \sum_{k>j} \beta_{ijk} x_i x_j x_k + \ldots + \sum_i \beta_{ii \ldots j} x_i^d, \tag{2.2}$$

where d is the highest order of the response surface model. Once the polynomial coefficient matrix $\mathbf{B} = [\beta_0, \beta_1, \ldots, \beta_{H-1}]$ is determined, the response surface model is created. The coefficient matrix is obtained by

$$\beta = (\mathbf{X}^{\top}\mathbf{X})^{-1}\mathbf{X}^{\top}\mathbf{Y}, \tag{2.3}$$

where $\mathbf{X}$ is the polynomial matrix of inputs, and $\mathbf{Y}$ is the output response value at the sample points. RSM is convenient to construct, requires few samples, and has some noise-filtering capability. Since the model itself is a weighted superposition of polynomials, it offers good transparency. However, its local fitting capability is relatively weak, and it may be inadequate for high-dimensional nonlinear problems.

2.1.1.2 Multivariate Adaptive Regression Splines (MARS)

MARS is a multidimensional, non-parametric, nonlinear regression modeling method [1]. It approximates unknown functional relationships using a combination of basis functions, defined as

$$\hat{f}(x) = \alpha_0 + \sum_{m=1}^{M} \alpha_m B_m(x), \tag{2.4}$$

where α_0 is a constant, M is the number of basis functions, $B_m(x)$ is the m-th basis function, and α_m is its weight coefficient. MARS effectively addresses the "Curse of Dimensionality" prevalent in multivariate regression. Its modeling process requires a few parameters to tune and relatively low computational effort. It can handle complex nonlinear problems. Similar to RSM, MARS models offer good transparency, but their construction often requires a larger number of sample points.

2.1.1.3 Radial Basis Functions (RBF)

RBF is a simple, exact interpolation method suitable for approximating deterministic computational simulation models. Its general form is

$$\hat{f}(x) = \sum_{i=1}^{s} \lambda_i \varphi(\|x - x^i\|). \tag{2.5}$$

Here, λ_i are weight coefficients and $\varphi(\cdot)$ is the radial basis function. RBF is a linear sum of basis functions centered on sample points. The choice of basis function and adjustment of weights significantly impact RBF's approximation performance. Its advantages lie in its simple structure and high accuracy, particularly suited for high-order, high-dimensional problems.

2.1.1.4 Kriging Model

Kriging is an interpolation-based approximation model grounded in variogram theory. It is a minimum-variance unbiased estimator, meaning it uses the change in prediction variance to measure changes in the design space while minimizing prediction error. Its standard model includes a global regression part and a stochastic process:

$$\hat{y}(x) = \mathbf{f}(x)^T \boldsymbol{\beta} + z(x), \tag{2.6}$$

where $\mathbf{f}$ is the vector of regression functions, $\boldsymbol{\beta}$ is the corresponding coefficient vector, and $z(x)$ is a realization of a stationary stochastic process. The Kriging model not only provides predicted values at estimation points but also estimates their variance, allowing assessment of prediction accuracy at those points. Kriging models can effectively approximate high-dimensional nonlinear problems, but their construction is relatively complex and computationally intensive [2].

2.1.1.5 Support Vector Regression (SVR)

SVR extends Support Vector Machines (SVM) to regression problems. Its prediction model is

$$\hat{y}(x) = \sum_{i=1}^{s} (\alpha_i - \alpha_i^*) \kappa(x, x^i) + b, \tag{2.7}$$

where $\kappa(x, x^i)$ is the kernel function, whose choice affects prediction accuracy. α_i, α_i^* are Lagrange multipliers, and b is the bias term. SVR handles nonlinear problems well and can accommodate stochastic errors. However, the settings of parameters like Lagrange multipliers and bias influence prediction accuracy and generalization ability, requiring additional computational optimization for optimal values.

2.1.2 Variable-Fidelity Surrogate Models

2.1.2.1 Related Concepts

For clarity in research and exposition, this section defines concepts and terms related to variable-fidelity surrogate models and their design optimization (some definitions refer to Fig. 2.1):

(1) **Design Variable**. Parameters describing the characteristics of an engineering product (e.g., geometry, material, stiffness, and damping), denoted x_i $(i = 1, 2, \ldots, n)$. Changing these values during optimization improves design performance.
(2) **Design Variable Vector**. The vector $\mathbf{x} = [x_1, \ldots, x_n]$ formed by n design variables. Any design vector represents a design alternative.
(3) **Design Space**. The real space formed by the coordinates of all design variables, denoted Ω.
(4) **Design Point**. A design alternative represented as a point in the design space, understood as the endpoint of the design vector (starting from the origin).
(5) **Sampling Point (Sample Point)**. A set of design points selected from the design space using a specific method or strategy. Variable-fidelity optimization requires selecting two sets of points from the design spaces of different fidelity models.
(6) **High-Fidelity Sample**. Design sample points within the design space of the high-fidelity model, denoted X_h, with corresponding responses Y_h. Low-fidelity samples are denoted X_l, with responses Y_l.
(7) **Feasible Design**. The design space contains all possible designs, but some are unacceptable (e.g., negative dimensions). A design satisfying all requirements imposed by constraints is feasible; otherwise, it is infeasible.
(8) **Constraint**. Restrictions during the design process. In engineering, constraints can be performance requirements (stability, deformation, stiffness, strength) or variable bounds (e.g., geometric limits).
(9) **Explicit Constraint**. Constraints expressible as explicit functional relationships between design variables. Constraints not expressible this way are **Implicit Constraints** (e.g., deformation, stress, frequency requiring FEA or dynamics simulation).

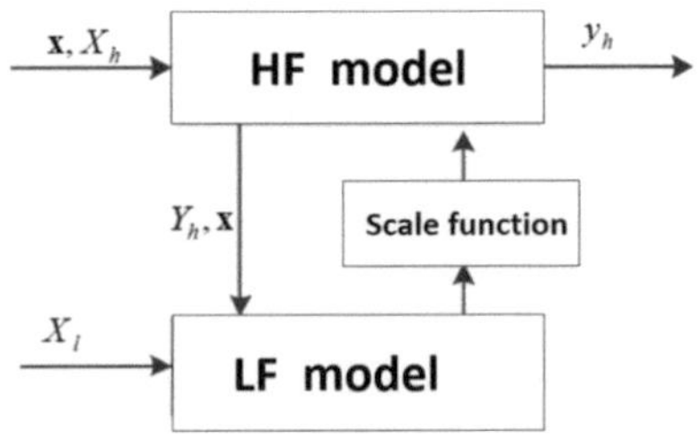

Fig. 2.1 Schematic of variable-fidelity surrogate model and design optimization

(10) **Feasible Region**. The region within the design space defined by constraints where feasible designs exist.
(11) **Objective Function**. A function representing one or more performance metrics used to evaluate design alternatives within the feasible domain. Optimizing this function yields better designs. Examples include volume, mass, time, energy, cost, stress, strain, or economic indicators. Multi-objective problems optimize several functions.
(12) **Optimal Design**. The feasible design that minimizes or maximizes the objective function value.

2.1.2.2 Scaling Functions

Additive and multiplicative scaling functions are common variable-fidelity models. Assuming locally that $F_h(x) = \beta(x) f_l(x)$ or $F_h(x) = f_l(x) + \gamma(x)$ holds, the scaling functions at a point x_n are

$$\text{Multiplicative Scaling} \quad \beta(x_n) = \frac{F_h(x_n)}{f_l(x_n)}, \tag{2.8}$$

$$\text{Additive Scaling} \quad \gamma(x_n) = F_h(x_n) - f_l(x_n). \tag{2.9}$$

Using Taylor expansion, $\tilde{\beta}(x) = \beta(x_n) + \nabla\beta(x_n)^T(x - x_n)$, $\tilde{\gamma}(x) = \gamma(x_n) + \nabla\gamma(x_n)^T(x - x_n)$. Gradients are computed as

$$\nabla\beta = \frac{\frac{\partial F_h}{\partial x_i} f_l - \frac{\partial f_l}{\partial x_i} F_h}{f_l^2}, \quad \nabla\gamma(x_n) = \nabla F_h(x_n) - \nabla f_l(x_n). \tag{2.10}$$

Thus, $F_h(x) \approx \tilde{\beta}(x) f_l(x)$, $F_h(x) \approx f_l(x) + \tilde{\gamma}(x)$. $\tilde{\beta}(x)$ and $\tilde{\gamma}(x)$ are the multiplicative and additive scaling functions, linking the HF model $F_h(x)$ and the LF model $f_l(x)$. The scaled $f_l(x)$ can replace the HF model in optimization to reduce cost. Replacing the Taylor expansion in $\tilde{\beta}(x)$ or $\tilde{\gamma}(x)$ with global models like Kriging, LSSVR, RSM, or RBF yields global variable-fidelity models valid over larger regions. Using Kriging as an example: Assume $F_h(x) = f_l(x) + \tilde{C}(x)$, where $\tilde{C}(x)$ is a Kriging scaling function. To build it, compute the discrepancy $C = \{c(x_1), c(x_2), \ldots, c(x_n)\}$:

$$c(x_i) = y^h(x_i) - y^l(x_i), \tag{2.11}$$

where $y^h(x_i)$ and $y^l(x_i)$ are HF and LF responses at x_i, respectively. Following the Kriging construction process, $\tilde{C}(x)$ is

$$\tilde{C}(x) = \hat{u} + \mathbf{r}^T\mathbf{R}^{-1}(\mathbf{Y}^h - \mathbf{Y}^l), \quad \hat{u} = (\mathbf{F}^T\mathbf{R}^{-1}\mathbf{F})^{-1}\mathbf{F}^T\mathbf{R}^{-1}(\mathbf{Y}^h - \mathbf{Y}^l), \tag{2.12}$$

where $\mathbf{R}(\cdot)$ is the correlation matrix, $\mathbf{r}^T(x) = [R(x, x_1), \ldots, R(x, x_n)]$, $\mathbf{Y}^h = [y^h(x_1), \ldots, y^h(x_{nh})]$, $\mathbf{Y}^l = [y^l(x_1), \ldots, y^l(x_{nl})]$, and $\mathbf{F} = [1, 1, \ldots, 1]^T$. The final Kriging-scaled variable-fidelity model is

$$\hat{F}_h(x) = f_l(x) + (\mathbf{F}^T\mathbf{R}^{-1}\mathbf{F})^{-1}\mathbf{F}^T\mathbf{R}^{-1}(\mathbf{Y}^h - \mathbf{Y}^l) + \mathbf{r}^T\mathbf{R}^{-1}(\mathbf{Y}^h - \mathbf{Y}^l). \quad (2.13)$$

Similarly, using RSM, RBF, LSSVR, etc., to approximate the HF-LF discrepancy builds other global scaling functions.

2.1.3 Ensemble Surrogate Models

No single surrogate model universally outperforms others. Combining models (Ensemble of Surrogates, EOS) via weighted summation leverages their strengths, improving prediction accuracy, robustness, and avoiding poor model choices. Weight determination is crucial. The ensemble model has the form:

$$\hat{y}^{\text{ens}}(x) = \sum_{i=1}^{N_i} \omega_i \hat{y}_i(x), \quad \sum_{i=1}^{N_i} \omega_i = 1, \quad (2.14)$$

where N_i is the number of models, $\hat{y}_i(x)$ is the prediction of the i-th model, and ω_i is its weight. Weighting methods fall into two categories: *Average Weighting* (constant ω_i based on global error metrics like global variance, heuristic methods, or optimal weighting) and *Pointwise Weighting* (ω_i varies per prediction point, e.g., based on k-nearest neighbors or spatial models) [3].

2.1.4 High-Dimensional Surrogate Models

Existing models (RSM, Kriging, RBF, VF) often perform poorly in high dimensions (≥ 10) due to the "Curse of Dimensionality." High-Dimensional Model Representation (HDMR) [4] offers a promising solution using a finite-order, hierarchical functional decomposition:

$$\begin{aligned} f(x) = f_0 &+ \sum_{i=1}^{d} f_i(x_i) + \sum_{1 \leq i < j \leq d} f_{ij}(x_i, x_j) + \ldots \\ &+ \sum_{1 \leq i_1 < i_2 < \ldots < i_r \leq d} f_{i_1 i_2 \ldots i_r}(x_{i_1}, x_{i_2}, \ldots, x_{i_r}) + \ldots + f_{12\ldots d}(x_1, x_2, \ldots, x_d), \end{aligned} \quad (2.15)$$

where x_i is the i-th dimension of $x \in \mathbb{R}^d$, $f(x)$ is the system response, and f_0 is the mean response. $f_i(x_i)$ is the first-order effect of x_i, $f_{ij}(x_i, x_j)$ is the second-order interaction effect, etc. HDMR exploits the statistical weakening of variable interactions as dimensionality increases.

2.1.5 *Validation and Evaluation Metrics for Surrogate Models*

Constructed models must be evaluated to ensure prediction accuracy meets requirements.

2.1.5.1 Validation Methods

Common methods are: (1) *Holdout Validation*, which randomly splits samples into training and validation sets; (2) *K-Fold Cross-Validation*, which randomly partitions samples into K folds, and uses $K - 1$ for training, 1 for validation, then repeats K times and averages the metrics; (3) *Leave-One-Out (LOO) Validation*, which uses one sample for validation and the rest for training; repeat for all samples; average metrics. LOO is a special case of K-fold where $K = N$.

2.1.5.2 Evaluation Metrics

This section outlines the key metrics used to evaluate model performance, focusing on four critical aspects: model accuracy, transparency, robustness, and computational efficiency. These metrics provide a comprehensive framework for assessing the effectiveness and practical utility of the models discussed in this monograph.

1. **Model Accuracy**: Quantifies the predictive performance of the model through standard regression metrics. The following metrics are employed:

$$R^2 = 1 - \frac{\sum_{i=1}^{N}(y_i - \hat{y}_i)^2}{\sum_{i=1}^{N}(y_i - \bar{y})^2}, \quad \text{(Coefficient of Determination)} \tag{2.16}$$

$$\text{RRMSE} = \frac{1}{\sigma}\sqrt{\frac{1}{N}\sum_{i=1}^{N}(y_i - \hat{y}_i)^2}, \quad \text{(Relative Root Mean Squared Error)} \tag{2.17}$$

$$\text{RAAE} = \frac{1}{N \cdot \sigma}\sum_{i=1}^{N}|y_i - \hat{y}_i|, \quad \text{(Relative Average Absolute Error)} \tag{2.18}$$

$$\text{RMAE} = \frac{1}{\sigma} \max_{i=1,\ldots,N} |y_i - \hat{y}_i|, \qquad \text{(Relative Maximum Absolute Error)} \tag{2.19}$$

$$\sigma = \sqrt{\frac{1}{N-1}\sum_{i=1}^{N}(y_i - \bar{y})^2}, \qquad \text{(Standard Deviation)} \tag{2.20}$$

where y_i is the observed value, $\hat{y}_i$ is the predicted value, $\bar{y}$ is the mean of observed values, and N is the sample size. The metrics RRMSE, RAAE, and R^2 assess global accuracy, with lower RRMSE and RAAE values and higher R^2 values (close to 1) indicating better performance. RMAE evaluates local accuracy, where lower values are preferable.

2. **Model Transparency**: Refers to the extent to which the model explicitly reveals its input-output relationships, facilitating interpretability and understanding by users and researchers.
3. **Model Robustness**: Measures the model's ability to maintain consistent accuracy across diverse conditions, such as variations in sample size, sampling methods, or problem types. Robustness is quantified by the standard deviation of accuracy metrics (RRMSE, RAAE, RMAE) across these conditions, with a lower standard deviation indicating greater robustness.
4. **Model Computational Efficiency**: Evaluates the computational resources required for model construction and prediction, typically measured by CPU time. Lower computational costs are desirable for practical implementation.

2.2 Design of Experiments

Design of Experiments (DOE) is a cornerstone of modern computer-aided product design, enabling efficient exploration of complex design spaces with limited computational resources [5]. DOE methods aim to maximize information gain from a minimal number of samples, which is critical for constructing accurate surrogate models in multidisciplinary design optimization. The choice of sample size and location significantly impacts model performance, as each analysis incurs computational costs. DOE strategies are broadly categorized into Classical Space-Filling Designs and Sequential Sampling Designs, each addressing different aspects of the sampling problem.

2.2.1 Classical Space-Filling Designs

Classical DOE methods employ "one-shot" sampling, generating all sample points simultaneously to cover the design space effectively. These methods are divided into two categories: designs focusing on margins and space-filling designs.

1. **Designs Focusing on Margins**: These prioritize placing sample points at the boundaries of the design space to capture extreme behaviors. Examples include Central Composite Design, Full/Fractional Factorial, Box-Behnken, and Plackett-Burman designs [6]. Such designs typically exhibit low random error but may introduce higher systematic error due to their focus on boundaries.
2. **Space-Filling Designs**: These aim for uniform distribution of points across the design space, minimizing clustering and ensuring comprehensive coverage. Key examples include:

   - *Maximin Design*: Maximizes the minimum Euclidean distance between any two points, promoting even spacing.
   - *Maximum Entropy Design*: Maximizes Shannon's entropy to optimize information gain [7].
   - *Uniform Design*: Ensures points are evenly distributed across the design space.
   - *Latin Hypercube Design (LHD)*: Partitions each of the d dimensions into n equal intervals, creating a grid of $d \times n$ cells. One point is randomly placed in each cell per dimension, ensuring non-collapsing projections (i.e., no two points share the same coordinate in any dimension). The sampling formula is

   $$X_{ij} = \frac{\pi_j(i) - U_{ij}}{n}, \quad i = 1, \ldots, n, \quad j = 1, \ldots, d, \tag{2.21}$$

     where n is the sample size, d is the number of dimensions, $\pi_j(i)$ is a random permutation of $\{1, \ldots, n\}$ for dimension j, and $U_{ij} \sim \mathcal{U}[0, 1]$ is a uniform random variable. LHD achieves superior accuracy compared to random sampling (e.g., Monte Carlo) with fewer points due to its space-filling and non-collapsing properties [8]. Figure 2.2 shows an LHD sample.
   - *Orthogonal Array Design*: Ensures orthogonality in projections, enhancing uniformity.

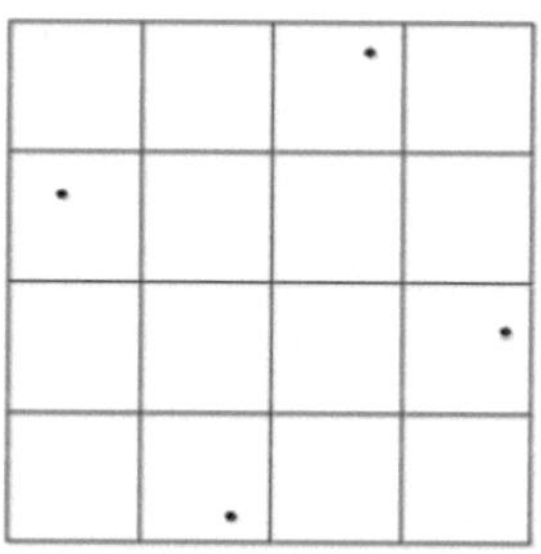

Fig. 2.2 Sample point distribution in LHD ($n = 4$)

2.2.2 Sequential Sampling Designs

Classical one-shot designs require prior knowledge of the optimal sample size, which is challenging for black-box problems with unknown response surfaces. Sequential sampling addresses this by iteratively adding points based on prior samples and model behavior, allowing adaptive exploration of the design space. Sequential sampling is divided into Uniform Sequential Sampling and Adaptive Sequential Sampling.

2.2.2.1 Uniform Sequential Sampling

Uniform sequential sampling extends space-filling designs (e.g., Maximin, Maximum Entropy, LHD) to a sequential framework. For instance, Sequential Optimal LHD solves an optimization problem to place new points that maximize the minimum distance to existing points while preserving projection constraints. As sample size increases, this process becomes computationally intensive, often requiring heuristic methods such as simulated annealing or evolutionary algorithms to ensure scalability.

2.2.2.2 Adaptive Sequential Sampling

Adaptive sequential sampling enhances efficiency for complex, multimodal functions by prioritizing sample placement in regions of high nonlinearity (e.g., peaks or valleys) while reducing sampling in flatter regions. New points are selected based on *infill criteria*, which balance two objectives:

- *Exploitation*: Prioritizing regions near the current optimum to refine local accuracy.
- *Exploration*: Targeting regions of high uncertainty or error to improve global coverage.

Table 2.1 summarizes common infill criteria, indicating their emphasis on exploitation versus exploration. Over-emphasizing exploitation risks missing the global optimum, while excessive exploration may slow convergence.

2.2.3 Experimental Design for Variable-Fidelity Surrogate Modeling

Variable-fidelity surrogate modeling leverages multiple models of differing computational costs and accuracies to efficiently explore complex design spaces in multidisciplinary design optimization. These models typically include a low-fidelity (LF) model, which is computationally inexpensive but less accurate, and a high-fidelity (HF) model, which is more accurate but computationally expensive. The experimental design for variable-fidelity surrogate modeling aims to strategically sample both

Table 2.1 Infill criteria for adaptive sequential sampling

Infill criterion	Exploitation	Exploration
Surrogate Management Framework (SMF) [9]	High	Low
Trust Region Method (TRM) [10, 11]	High	Low
Minimizing the Predictor [12]	High	Low
Expected Improvement (EI) [13, 14]	High	Low
Generalized Expected Improvement [15]	High	Low
Lower Confidence Bound (LCB) [16, 17]	High	Low
Probability of Improvement (PI) [16, 18]	High	Low
Maximum Predicted Variance [17]	Low	High

LF and HF models to construct accurate surrogate models while minimizing computational costs. This subsection outlines the principles governing variable-fidelity experimental design, with a focus on nested designs that facilitate discrepancy modeling between fidelity levels. The discussion lays the foundation for subsequent chapters, which present two distinct nested Latin Hypercube Design (LHD) methods for constructing variable-fidelity surrogate (VFS) models.

The design of experiments for variable-fidelity modeling adheres to three fundamental principles [19]:

1. **Economy**: The number of HF samples, $N(D_h)$, is significantly smaller than the number of LF samples, $N(D_l)$, i.e., $N(D_h) < N(D_l)$. This principle exploits the computational efficiency of LF models, which can be evaluated at a larger number of points to broadly explore the design space, while HF evaluations are reserved for critical regions to refine accuracy.
2. **Nested Relationship**: Traditionally, a nested relationship, where the HF sample set D_h is a subset of the LF sample set D_l (i.e., $D_h \subset D_l$), was preferred. Nested points, evaluated by both LF and HF models, enable direct discrepancy modeling (e.g., additive or multiplicative corrections). However, modern methods like transfer learning eliminate the need for nested structures by leveraging knowledge transfer from LF to HF models, allowing independent sampling of D_h and D_l. This flexibility enhances scalability but requires sophisticated alignment techniques to ensure compatibility between fidelity levels.
3. **Space-Filling**: Both D_h and D_l must exhibit space-filling properties to ensure comprehensive coverage of their respective design spaces, as the optimal design region is typically unknown a priori. Space-filling is quantified by metrics such as the minimum Euclidean distance between points or the uniformity of point projections [20]. Effective space-filling ensures that the surrogate model captures the global behavior of the underlying system, reducing the risk of missing critical regions.

To illustrate these concepts, consider a two-dimensional design space. In a *nested design* (see Fig. 2.3a), the LF sample set D_l comprises 16 points distributed via an

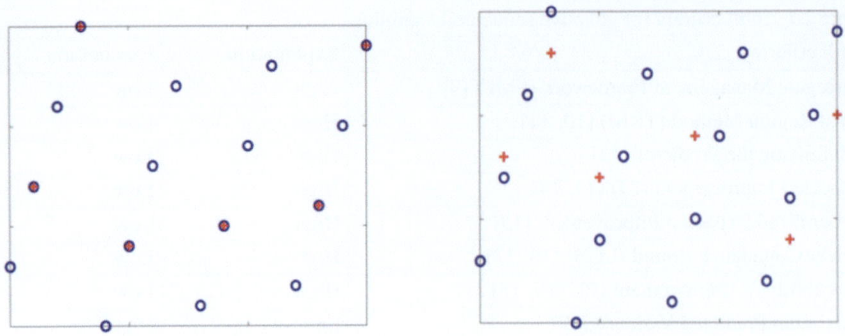

(a) Nested Design (b) Non-Nested Design

Fig. 2.3 Nested versus non-nested experimental designs

LHD to ensure space-filling properties. The HF sample set D_h, with 6 points, is a subset of D_l, selected to optimize both space-filling and discrepancy modeling. These shared points facilitate direct comparison of LF and HF outputs. In a *non-nested design* (see Fig. 2.3b), D_h and D_l are sampled independently, such that HF points may not coincide with LF points. For example, transfer learning approaches might use 16 LF points to train a base model and 6 distinct HF points to fine-tune it, relying on model-based alignment rather than shared evaluations. Non-nested designs offer greater flexibility in sample placement but may complicate discrepancy modeling unless robust transfer mechanisms are employed.

Transfer learning represents a paradigm shift in variable-fidelity modeling by relaxing the nested structure constraint. In this approach, an LF model is pre-trained on a large set of samples (D_l) to capture general trends, and its knowledge is transferred to an HF model through techniques such as fine-tuning or feature sharing. The HF model is then trained on a smaller, potentially independent set of samples (D_h). This method reduces the dependency on shared evaluation points, allowing greater freedom in experimental design. However, it introduces challenges in ensuring alignment between fidelity levels, often requiring advanced machine learning techniques to mitigate domain shifts. Transfer learning is particularly advantageous when LF and HF models share structural similarities or when computational budgets preclude extensive HF sampling.

The principles outlined above underpin the experimental designs explored in the following chapters. Chapters 3 and 4 present two nested LHD methods for variable-fidelity surrogate modeling. The first method achieves the fast nested LHD via translation propagation. The second method optimizes the space-filling properties of D_h within a nested D_l, ensuring robust discrepancy modeling. Additionally, a transfer learning approach, which eschews the nested structure, is introduced in Chap. 7 to leverage non-nested sampling for scalable VFS construction. These methods collectively address the trade-offs between computational efficiency, model accuracy, and design flexibility in multidisciplinary optimization.

Designing experiments for variable-fidelity surrogate modeling requires balancing sample sizes, sampling relationships, and space-filling objectives. Nested designs offer simplicity in discrepancy modeling but constrain sample placement. Non-nested approaches, such as transfer learning, provide flexibility but demand robust alignment strategies. The choice of $N(D_h)$ relative to $N(D_l)$ depends on the computational budget and the fidelity gap between models. The methods discussed in subsequent chapters provide practical frameworks for navigating these challenges.

References

1. Friedman JH et al (1991) Multivariate adaptive regression splines. Ann Stat 19(1):1–67
2. Han Z (2016) Kriging surrogate model and its application to design optimization: a review of recent progress. Acta Aeronautica et Astronautica Sinica 37(11):3197–3225
3. Jiang Z (2016) Research on ensemble of surrogates and sequential sampling strategies for product design optimization. PhD thesis. Huazhong University of Science and Technology
4. Sobol IM (2001) Global sensitivity indices for nonlinear mathematical models and their Monte Carlo estimates. Math Comput Simul 55(1–3):271–280
5. Durakovic B (2017) Design of experiments application, concepts, examples: state of the art. Period Eng Nat Sci 5(3):421–439
6. Liu H (2016) The research and application of metamodel-based optimization. PhD thesis. Dalian University of Technology
7. Wynn HP (2004) Maximum entropy sampling and general equivalence theory. In: mODa 7—advances in model-oriented design and analysis: proceedings of the 7th international workshop on model-oriented design and analysis held in Heeze, The Netherlands, June 14–18, 2004. Springer, pp 211–218
8. Levy S, Steinberg DM (2010) Computer experiments: a review. AStA Adv Stat Anal 94:311–324
9. Booker AJ, Dennis JE, Frank PD et al (1999) A rigorous framework for optimization of expensive functions by surrogates. Struct Opt 17:1–13
10. Rodrıéguez JF, Renaud JE, Wujek BA et al (2000) Trust region model management in multidisciplinary design optimization. J Comput Appl Math 124(1–2):139–154
11. Kim DW, Lee J (2010) An improvement of Kriging based sequential approximate optimization method via extended use of design of experiments. Eng Opt 42(12):1133–1149
12. Ziegel ER (2003) The elements of statistical learning
13. Jones DR, Schonlau M, Welch WJ (1998) Efficient global optimization of expensive black-box functions. J Global Opt 13:455–492
14. Huang Y-C, Chan K-Y (2010) A modified efficient global optimization algorithm for maximal reliability in a probabilistic constrained space
15. Wolfgang P, Tobias W, Markus V (2008) Clustered multiple generalized expected improvement: a novel infill sampling criterion for surrogate models. In: IEEE congress on evolutionary computation (IEEE world congress on computational intelligence). IEEE, 2008, pp 3515–3522
16. Jones DR (2001) A taxonomy of global optimization methods based on response surfaces. J Glob Opt 21:345–383
17. Sasena MJ, Papalambros P, Goovaerts P (2002) Exploration of metamodeling sampling criteria for constrained global optimization. Eng Opt 34(3):263–278
18. Forrester AIJ, Keane AJ (2009) Recent advances in surrogate-based optimization. Progr Aerosp Sci 45(1–3):50–79
19. Qian PZG, Tang B, Wu CFJ (2009) Nested space-filling designs for computer experiments with two levels of accuracy. Stat Sin 287–300
20. Crombecq K, Laermans E, Dhaene T (2011) Efficient space-filling and non-collapsing sequential design strategies for simulation-based modeling. Eur J Oper Res 214(3):683–696

Chapter 3
Fast Nested Latin Hypercube Design via Translation Propagation

In modern product design, experimental design is a key technology in computer-aided design. A well-designed experiment allows designers to gather more information about the design space with a limited number of sample points. In the process of building an experimental design is an essential step. To obtain the response values of the objective function at these points, the corresponding analysis models must be run at these points. Therefore, the sample size directly reflects the number of runs of the analysis model, which is related to sampling time and cost. In addition, studies have shown that the position of the sample points and the sample size also affect the performance of the constructed surrogate models. In the construction of variable-fidelity surrogate models, sample points come from design spaces corresponding to different fidelity analysis models, making the sample configuration process more complex. Thus, in-depth research is needed for variable-fidelity surrogate models.

The previous chapter introduced three fundamental principles that experimental designs for variable-fidelity surrogate models should possess: economicity, space-filling, and nested relationships. This chapter presents a sample distribution method for variable-fidelity surrogate models, namely, a nested Latin hypercube design based on translation propagation.

3.1 Latin Hypercube Design Based on Translation Propagation

This section proposes a nested Latin hypercube design method based on the translation propagation approach. The main strategy is to first construct a simple seed design with nested relationships, and then translate and propagate the nested seed across the entire design space according to certain rules to obtain the final nested design. The final design possesses the characteristics of Latin hypercube design; hence, it is also referred to as nested Latin hypercube design.

J. Yi et al., *Variable-Fidelity Surrogate*, Engineering Applications of Computational Methods 25, https://doi.org/10.1007/978-981-95-5527-7_3

3.1.1 Nested Seed Design

The basic idea of nested Latin hypercube design is to translate and propagate nested seed designs across the entire design space. Therefore, nested seed design is the foundation of experimental design methods for variable-fidelity surrogate models. This paper presents different nested seed designs, as shown in Figs. 3.1, 3.2 and 3.3. In these figures, red circles represent HF sample points, and blue points represent LF sample points. All red points are covered by blue points, meaning that sample points for high-fidelity analysis models are also used for LF analysis model sampling, i.e., these seed designs have nested relationships. Figure 3.1 illustrates various two-dimensional nested seed designs, such as 2 nested 1, 3 nested 1, 3 nested 2, 4 nested 1, and 4 nested 2 seeds (shown in Fig. 3.1a–e).

In addition to the two-dimensional seed designs shown in Fig. 3.1, higher-dimensional nested seeds have also been developed. For example, Fig. 3.2 shows three-dimensional seed designs, including 3 nested 1, 3 nested 2, 4 nested 1, and 4 nested 2 seed designs (shown in Fig. 3.2a–d). Figure 3.3 presents a four-dimensional 4 nested 2 seed design.

3.1.2 Design Space Partitioning

Assuming an n_v-dimensional design space, to utilize an n_{s1} nested n_{s2} seed (where n_{s1} is the number of LF sample points in the seed design, and n_{s2} is the number of HF sample points) to design a nested Latin hypercube with n_{p1} LF sample points and n_{p2} HF sample points, the main idea is to partition the design space into several identical "blocks." The nested seeds are then translated and propagated across these blocks. For example, in a two-dimensional design space, using an $n_{s1} = 2$, $n_{s2} = 1$ seed (as shown in Fig. 3.1) to design an 18 nested 9 nested Latin hypercube, it is first necessary to determine the number of blocks, the number of blocks between, and the initial sample size.

A nested design involves two layers. Considering a single layer, the LF layer requires n_{b1} blocks as shown in Eq. 3.1, and the HF layer requires n_{b2} blocks. In the nested design, the larger value between the two is taken, as shown in Eq. 3.2:

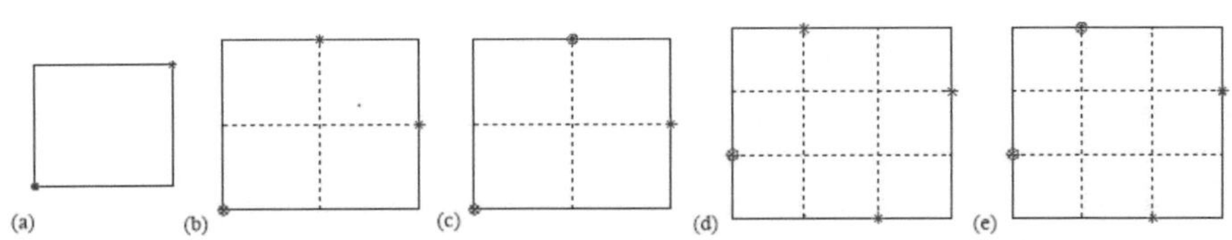

Fig. 3.1 Various two-dimensional nested seed designs

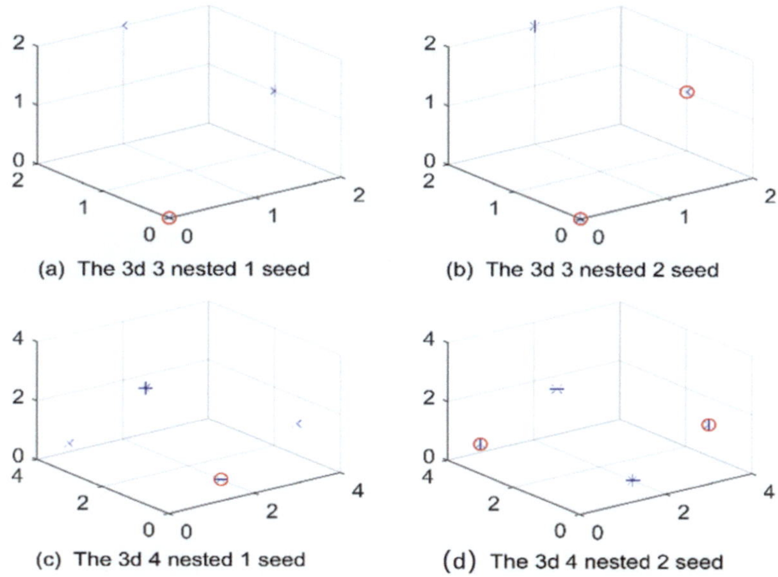

Fig. 3.2 Three-dimensional seed design

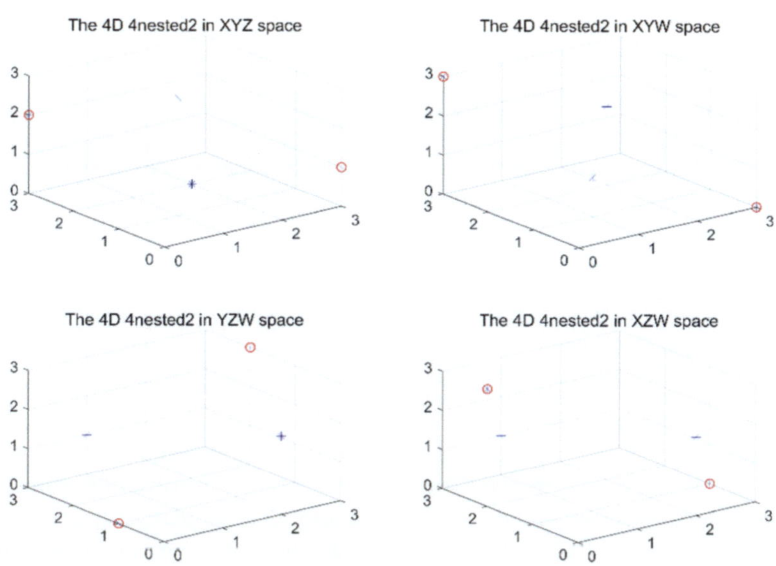

Fig. 3.3 Four-dimensional 4 nested 2 seed design

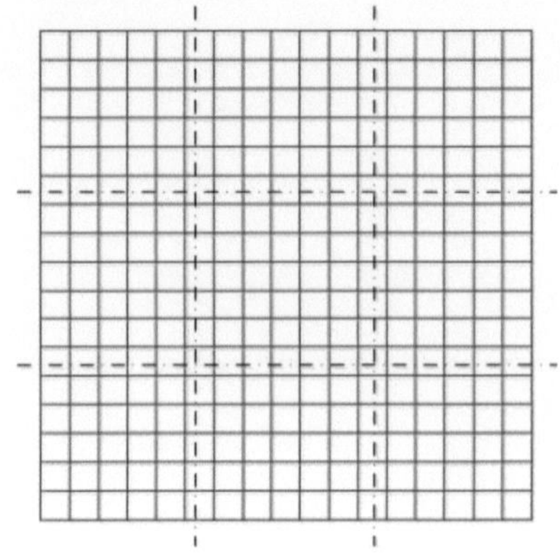

Fig. 3.4 Nested LHD design space partitioning

$$n_{b1} = \frac{n_{p1}}{n_{s1}}, \quad n_{b2} = \frac{n_{p2}}{n_{s2}} \tag{3.1}$$

$$n_b = \max(n_{b1}, n_{b2}) \tag{3.2}$$

From Eq. 3.2, the number of intervals along each dimension can be obtained as follows:

$$n_d = (n_b)^{1/n_v}, \tag{3.3}$$

Thus, the design space is partitioned into n_b identical blocks. As shown in Fig. 3.4, the design space for 18 nested 9 is divided into 9 blocks, with each dimension divided into 3 intervals, each interval having six levels.

3.1.3 Initial Design

The seed is filled into each block along the dimension, as shown in Fig. 3.5. First, as shown in Fig. 3.5a, the nested seed is appropriately stretched and placed into the first block generated in step two. The stretching process involves inserting levels between two consecutive sample points along the same dimension (the number of inserted levels is $n_d - 1$), as shown in Fig. 3.6. The purpose of this is to enhance the space-filling property of the seed design. Then, this block containing the seed design is moved along a certain dimension to the next block (moving n_{p1}/n_d^2 levels each time). With each move, n_{s1} LF points and n_{s2} HF points are added to the entire experimental design. Figure 3.5b shows the first move (in the horizontal direction). In addition to translation, there is an offset in the other dimension (the vertical direction in the figure). The seed design is translated and propagated in this manner until all intervals of a certain dimension are filled, as shown in Fig. 3.5a–c. Then, all design

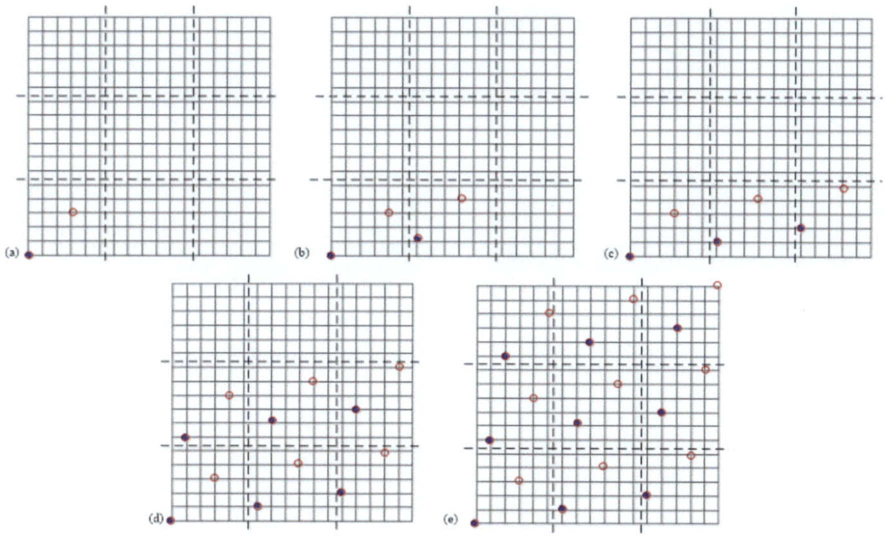

Fig. 3.5 Generation process of nested LHD based on translation propagation

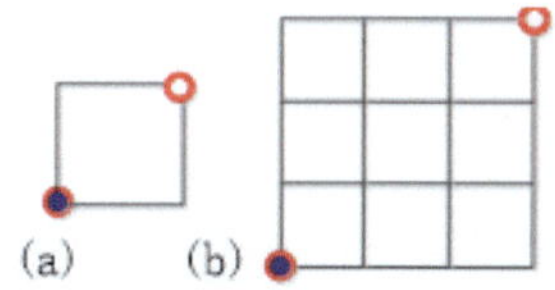

Fig. 3.6 Nested seed stretching

points along this dimension (as shown in Fig. 3.5c) are treated as a new seed and moved along the other dimension, as shown in Fig. 3.5d–e. This process is repeated, and after moving along each dimension, the final design is obtained.

The greatest advantage of this experimental design generation method is that it requires no computation. The entire process can be viewed as a simple replication and propagation of the nested seed design within an n_b-dimensional design space, resulting in minimal computational effort and ease of operation. However, this method can only produce experimental design schemes when n_{b1} and n_{b2} in Eq. 3.1 are equal and n_d in Eq. 3.3 is an integer. To generate nested designs of any sample size for variable-fidelity approximation, the following section will outline the corresponding strategy.

3.1.4 Removal of Excess Points

In practical situations, the number of intervals must be an integer. When n_d in Eq. 3.3 is not an integer, the actual number of intervals per dimension is

$$n_d^* = \lceil n_d \rceil , \tag{3.4}$$

where $\lceil\rceil$ denotes the ceiling function.

For such cases, the strategy to generate nested designs of any sample size using the translation propagation approach is as follows: first, produce a nested design with more samples than required, and then remove the excess sample points according to certain principles.

By Eq. 3.4, the actual number of blocks in the design space is

$$n_b^* = (n_a^*)^{n_v}. \tag{3.5}$$

Using the method from Sect. 3.1.2, the number of LF and HF points in the generated nested Latin hypercube design are

$$n_{\text{pin1}} = n_{i1} \times n_b^*, n_{\text{pin2}} = n_{i2} \times n_b^*. \tag{3.6}$$

If $n_{\text{pin1}} = n_{p1}$ and $n_{\text{pin2}} = n_{p2}$, then the case described in Sect. 3.1.2 applies, and no subsequent removal of excess points is necessary. Otherwise, excess sample points must be removed. Since the experimental design for variable-fidelity approximation is a nested two-layer design, removing points from the LF design may involve removing points from the HF sample locations or points outside the HF sample locations. Therefore, removing excess points may alter the nested relationship between HF and LF samples. To address this, the following method is employed for removing excess points.

When the number of points to be removed from the LF sample exceeds that from the HF sample: (1) first remove the $n_{\text{pin2}} - n_{p2}$ points farthest from the center of the HF point set, and then (2) for the LF sample points, first remove the $n_{\text{pin2}} - n_{p2}$ LF points corresponding to the removed HF points, and subsequently remove the remaining $n_{\text{pin1}} - n_{\text{pl}} - n_{\text{pin2}} + n_{p2}$ LF sample points from the set of LF samples not belonging to the HF samples. Conversely, when the number of points to be removed from the LF sample is less than that from the HF sample: (1) first remove $n_{\text{pin1}} - n_{p1}$ LF sample points from the nested samples to complete the removal of LF points, and then (2) for the HF sample points, first remove the $n_{\text{pin1}} - n_{p1}$ HF points corresponding to the removed nested samples, and then remove an additional $n_{\text{pin2}} - n_{p2} - n_{\text{pin1}} + n_{p1}$ points from the remaining HF sample set. When selecting points to be removed, the principle of removing points farthest from the design center is followed. The distance calculation is performed after the initial design is generated, that is, before any removal of nested sample points. All HF and LF points to be removed must be determined before the removal process. Once the excess

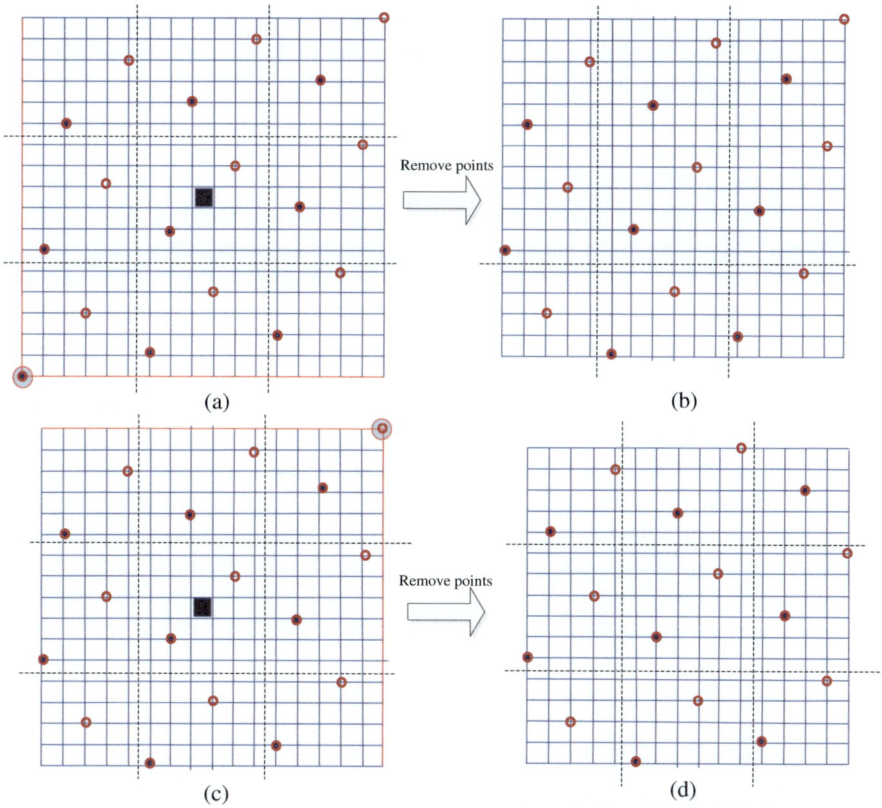

Fig. 3.7 Generation process of 16 nested 8 Latin hypercube design from 18 nested 9

points are removed, the final experimental design must be stretched to the actual design space. This stretching is achieved by removing the levels corresponding to the removed points (for LF points only or nested points) after each removal.

Figure 3.7 illustrates the generation process of a 16-nested-8 Latin hypercube design. According to Eqs. 3.5 and 3.6, the initial design block count is calculated as 9, with 18 LF initial sample points and 9 HF initial sample points. First, the initial design of 18 nested 9 is generated using the method from Sect. 3.1.3, as shown in Fig. 3.5. Since it is necessary to remove 2 LF points and 1 HF point, one HF point is first removed, as shown in Fig. 3.7, where the gray square represents the design center, and the red-circled points indicate the HF and LF points to be removed. Then, the level corresponding to the removed point is eliminated (indicated by the red border in the figure), thereby stretching the design after removal to the corresponding design space, as shown in Fig. 3.7b. At this point, the HF sample size meets the requirement, and only one LF sample point needs to be removed from the set of LF samples not belonging to the HF samples. Following the principle of removing the point farthest from the design center, the LF point at the upper right corner is removed along with its corresponding level (as shown in Fig. 3.7c), resulting in the final 16 nested 8 nested Latin hypercube design, as depicted in Fig. 3.7d.

If the block count calculated according to Eq. 3.4 is equal to either the HF or LF block count from Eq. 3.1, then only one type of sample point (either HF or LF) needs to be removed. If $n_{b1} > n_{b2}$ and $n_b^* = n_{b1}$, then only the excess HF sample points need to be removed. In this case, the HF points farthest from the design center are removed according to the aforementioned principle, without the need to remove related levels or stretch the design space. If $n_{b1} < n_{b2}$ and $n_b^* = n_{b2}$, then only the excess LF sample points need to be removed. However, the levels corresponding to the removed LF sample points must also be eliminated, followed by stretching the design space.

Through the aforementioned process, the final design ensures that there is only one experimental design point in each interval of every dimension, thereby satisfying the fundamental characteristics of Latin hypercube design. Consequently, this method is referred to as a nested Latin hypercube design. The translation propagation approach enables the rapid generation of nested Latin hypercube designs of any size, automatically fulfilling the first two principles for variable-fidelity approximation outlined in Chap. 2 (economicity and nested relationships) while also ensuring a certain degree of space-filling. For nested designs of the same size, different seeds yield different designs. However, it is not possible to predict in advance which seed will produce the highest-quality design (primarily in terms of space-filling).

3.1.5 Nested Seed Selection

Different seeds lead to different designs when translated and propagated. For instance, using the 2 nested 1, 3 nested 1, 3 nested 2, and 4 nested 1 seeds shown in Fig. 3.1a–d results in different 16 nested 8 designs, as depicted in Fig. 3.8a–d.

Since it is not possible to determine in advance which seed will yield the best space-filling experimental design, multiple nested seeds can be employed to generate designs, followed by calculating the minimum Euclidean distance between sample points in each design to select the one with the best space-filling performance. To evaluate the space-filling quality of an experimental design, the minimum Euclidean distance between sample points can be calculated using the following formula:

$$\min_{p_i, p_j \in P} \sqrt{\sum_{k=1}^{n_v} \left| p_i^k - p_j^k \right|^2}. \tag{3.7}$$

In the context of nested experimental designs for variable-fidelity approximation, the acquisition cost of HF sample points is significantly higher than that of LF points. For simplicity, this paper uses the space-filling quality of HF samples to assess the

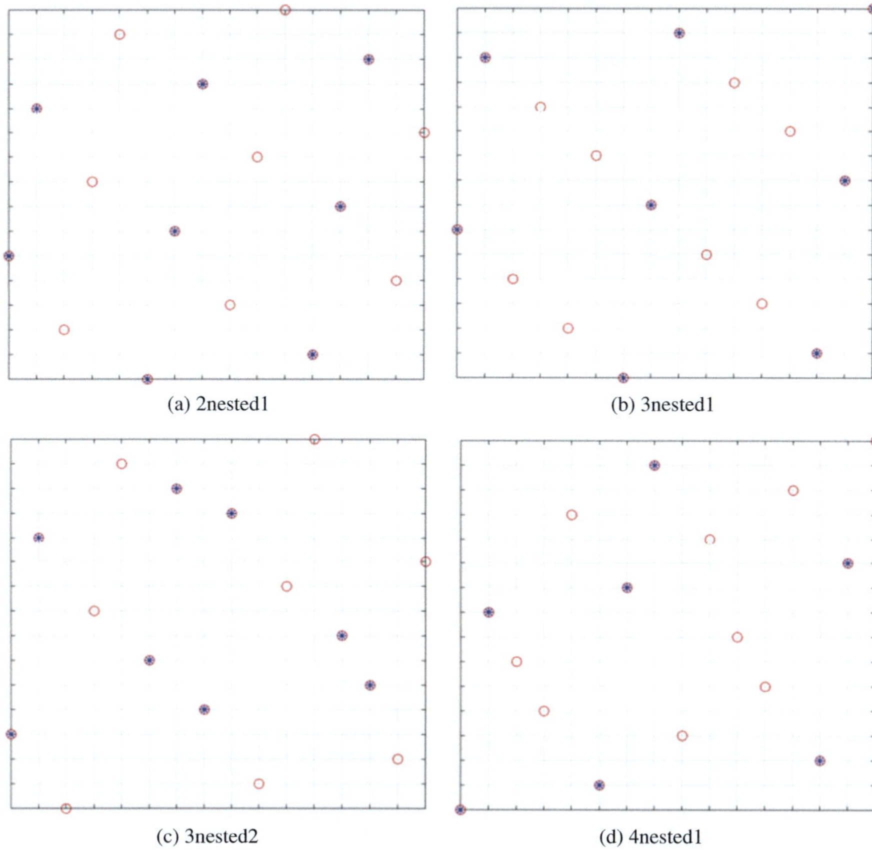

Fig. 3.8 18 nested 8 designs from different seeds selection

overall quality of the nested design. The minimum Euclidean distance for HF samples is calculated as follows:

$$\min_{p_i, p_j \in p} \left\| \sum_{k=1}^{n_v} (p_i^k - p_j^k)^2 \right\|. \tag{3.8}$$

Among several designs, the one with the maximum HF sample minimum distance (calculated using Eq. 3.8) is considered to have the best space-filling performance.

3.1.6 Algorithm Summary

In summary, the basic flowchart for constructing nested Latin hypercube designs using the translation propagation approach is presented in Fig. 3.9. First, based on the given HF and LF sample sizes, design space dimensions, nested seed design, and the number of HF and LF sample points in the seed, the number of intervals

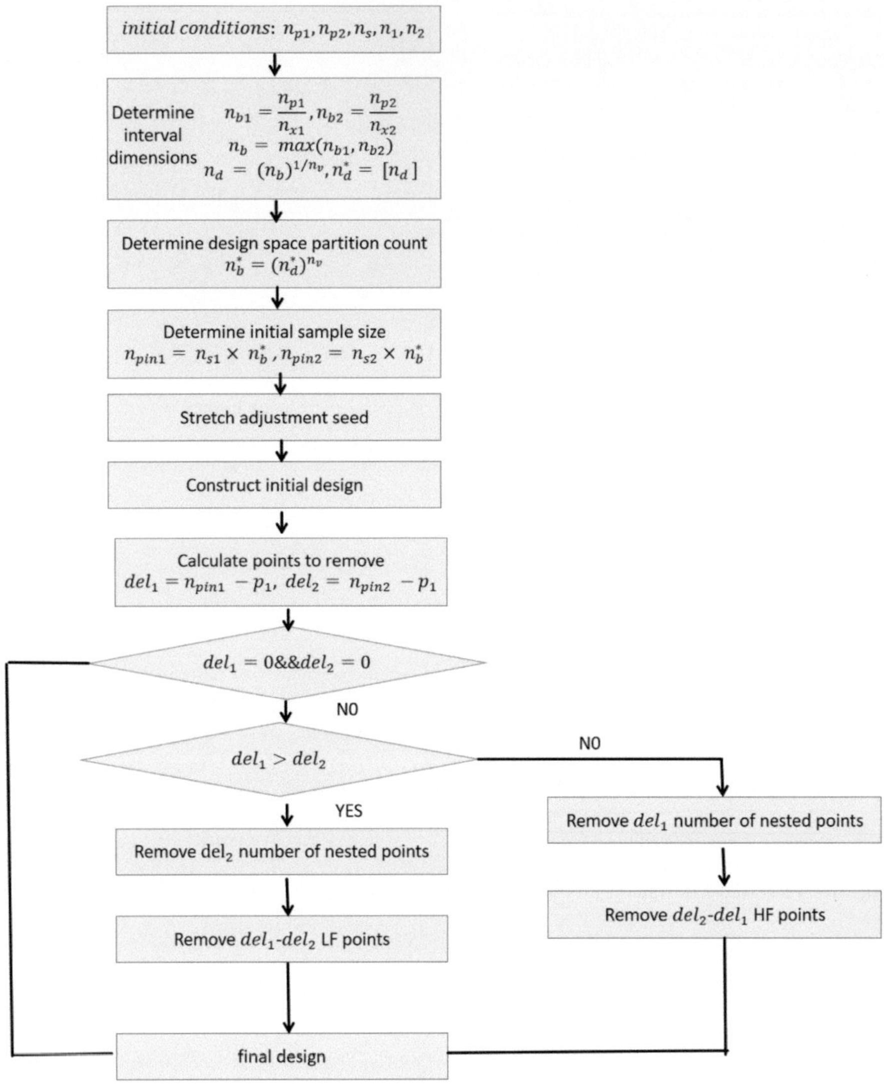

Fig. 3.9 Basic flowchart of nested Latin hypercube design based on translation propagation

per dimension is calculated using Eqs. 3.1 to 3.4. Then, the number of blocks in the design space is determined using Eq. 3.5. Subsequently, the initial design sample size is determined based on the block count. The nested seed design is stretched and propagated across the design space to obtain the initial nested Latin hypercube design. Next, the number of sample points to be removed from the initial design is calculated. If neither HF nor LF samples require removal, the initial design is the final design. Otherwise, excess HF and LF sample points must be removed. If the number of LF points to be removed exceeds that of HF points, first remove del_2 HF points and the corresponding LF sample points (i.e., del_2 nested points), and then remove the remaining $del_1 - del_2$ LF sample points from the set of LF samples not belonging to HF points. If the number of HF sample points to be removed exceeds that of LF points, first remove del_1 nested points (removing del_1 HF and LF sample points), and then remove $del_2 - del_1$ HF sample points from the remaining nested points. After removal, the final design is obtained.

3.2 Numerical Test Cases

3.2.1 Numerical Test Cases

This subsection employs the following functions to evaluate the performance of nested Latin hypercube designs. For each function, nested and non-nested Latin hypercube designs of the same size are generated. Different variable-fidelity surrogate models are then constructed based on these designs, and the prediction accuracy of the models is calculated to validate the effectiveness and specificity of the proposed nested Latin hypercube experimental design method.

Six-Hump Camel-Back Function (SC):

$$\begin{aligned} sc(x_1, x_2) &= 4x_1^2 - 2.1x_1^4 + \frac{x_1^6}{3} + x_1x_2 \\ &\quad - 4x_2^2 + 4x_2^4; \\ y_h &= sc(x_1, x_2); \\ y_l &= sc(0.7x_1, 0.7x_1) + x_1x_2 - 65; \\ x_1 &\in [-2, 2], \quad x_2 \in [-2, 2]. \end{aligned} \tag{3.9}$$

Modified Branin Function (MBR):

$$\begin{aligned} f(x_1, x_2) &= 10 + [x_2 - 5.1 \times x_1^2/(4\pi^2) + 5x_1/\pi - 6]^2 + 10\cos(x_1)[1 - 1/(8\pi)]; \\ y_h &= f(x_1, x_2) - 22.5x_2; \\ y_l &= f(0.7x_1, 0.7x_2) - 15.75x_2 + 20(0.9 + x_1)^2 - 50; \\ x_1 &\in [-5, 10], x_2 \in [0, 15]. \end{aligned} \tag{3.10}$$

Booth Function:

$$\begin{aligned} f(x_1, x_2) &= (x_1 + 2x_2 - 7)^2 + (2x_1 + x_2 - 5)^2 \\ x_{1,2} &\in [-10, 10] \\ y_h &= f(x_1, x_2) \\ y_l &= f(0.4x_1, x_2) + 1.7x_1x_2 - x_1 + 2x_2. \end{aligned} \tag{3.11}$$

Bohachevsky Function:

$$\begin{aligned} f(x_1, x_2) &= x_1^2 + 2x_2^2 - 0.3\cos(3\pi x_1) - 0.4\cos(4\pi x_2) + 0.7 \\ x_{1,2} &\in [-100, 100] \\ y_h &= f(x_1, x_2) \\ y_l &= f(0.7x_1, x_2) + x_1 x_2 - 12. \end{aligned} \tag{3.12}$$

Himmelblau Function (HM):

$$\begin{aligned} f(x_1, x_2) &= (x_1^2 + x_2 - 11)^2 + (x_2^2 + x_1 - 7)^2; \\ y_h &= f(x_1, x_2); \\ y_l &= f(0.5x_1, 0.8x_2) + x_2^3 - (x_1 + 1)^2; \\ x_{1,2} &\in [-3, 3]. \end{aligned} \tag{3.13}$$

3D Rosenbrock Function:

$$\begin{aligned} f(x_1, x_2, x_3) &= \sum_{i=1}^{2} \left[100(x_i^2 - x_{i+1})^2 + (x_i - 1)^2\right] \\ y_h &= f(x_1, x_2, x_3) \\ y_l &= 90(x_1^2 - x_2)^2 + 1.1(x_3 - 1)^2 + 100(x_2^2 - x_3)^2 + (x_2 - 1)^2 + \varepsilon \\ x_1 &\in [-2.048, 2.048], \quad x_2 \in [-2.048, 2.048], \end{aligned} \tag{3.14}$$

where ε is normally distributed noise with mean 0 and variance σ^2, i.e., $\varepsilon(x_1, x_2) \sim N(0, \sigma^2)$, where $\sigma^2 = \sigma_s^2/200$, and σ_s^2 represents the variance of the smooth part of the 3D Rosenbrock function.

Borehole Model:

$$\begin{aligned} f_{\text{borehole}} &= \frac{2\pi x_3[x_4 - x_6]}{\ln(x_2/x_1)[1 + \frac{2x_7 x_4}{\ln(x_2/x_1)x_1^2 x_8} + \frac{x_3}{x_5}]} \\ y_h &= f_{\text{borehole}}(x); \\ y_l &= 0.4 f_{\text{borehole}}(x) + 0.07x_1^2 x_8 + x_1 x_7/x_3 + x_1 x_6/x_2 + x_1^2 x_4; \\ x_1 &\in [0.05, 0.15], x_2 \in [100, 50000], \\ x_3 &\in [63070, 115600], x_4 \in [990, 1110], \\ x_5 &\in [63.1, 116], x_6 \in [700, 820], \\ x_7 &\in [1120, 1680], x_8 \in [9855, 12045]. \end{aligned} \tag{3.15}$$

The above is a borehole flow model used by many scholars to compare different computer experiment analysis methods [1].

3.2.2 Variable-Fidelity Surrogate Models

This section combines several different variable-fidelity surrogate models to evaluate the proposed experimental design method in the context of variable-fidelity surrogate model construction. Specifically, variable-fidelity surrogate models are constructed based on nested Latin hypercube designs and non-nested Latin hypercube designs, and the performance of the constructed models is compared. The main models used include bridge functions based on LSSVR, Kriging, RBF, and RSM, as well as single-fidelity surrogate models such as Kriging, RBF, and RSM.

3.2.3 Sample Design

Different sample sizes are employed to explore the impact of the proposed experimental design method on various surrogate models. Specifically, designs of different sizes, such as 50 nested 10, 50 nested 20, 50 nested 30, and 50 nested 40, are considered. Non-nested Latin hypercube designs of the same size are also constructed for comparison.

3.2.4 Test Results

Figure 3.10a–d, respectively, display the 50 nested 10, 50 nested 20, 50 nested 30, and 50 nested 40 nested Latin hypercube designs determined using the proposed method for the SC test case.

Tables 3.1 and 3.2 present the prediction accuracy of different variable-fidelity surrogate models under various experimental design methods, with three prediction accuracy criteria: Root Mean Square Error (RMSE), Maximum Absolute Error (MAE), and R-squared value. The "Nested" column represents the prediction errors of variable-fidelity surrogate models constructed using the nested Latin hypercube design based on translation propagation, while the "Non-nested" column represents the prediction errors under the non-nested Latin hypercube design. "DMF-LSSVR" denotes the bridge function based on LSSVR (i.e., using LSSVR to construct the surrogate model of the difference between LF and HF models). Similarly, "DMF-Kriging" [2, 3], "DMF-RSM" [4], and "DMF-RBF" refer to bridge functions based on Kriging, RSM, and RBF, respectively. These models utilize information from both HF and LF models during construction. In contrast, single-fidelity surrogate models such as Kriging, RBF, and RSM only use information from the high-fidelity model. From Table 3.1, it can be observed that as the number of sample points increases, the RMSE and MAE values of all surrogate models (including variable-fidelity models) decrease, while the R-squared value increases. This indicates that the accuracy of the constructed surrogate models improves with an increasing number of sample points.

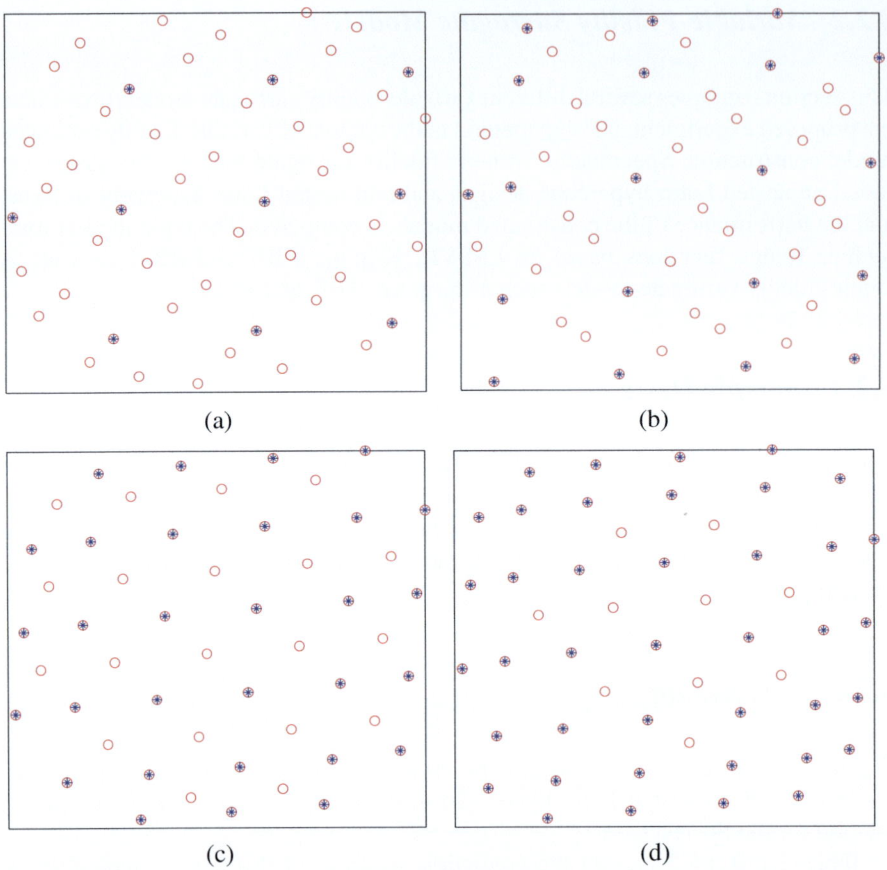

Fig. 3.10 Nested Latin hypercube designs of different sizes

Horizontally, for the same sample size, the RMSE and MAE values in the "Nested" column of the same variable-fidelity surrogate model are smaller than those in the "Non-nested" column, and the R Square value in the "Nested" column is larger than that in the "Non-nested" column (except for the Booth function with 10 sample points, where the "Non-nested" DMF-LSSVR has better precision than the "Nested" DMF-LSSVR). This indicates that models established on the nested Latin hypercube design have higher prediction accuracy. To better illustrate this, taking the deep hole model as an example, Fig. 3.11a–c, respectively, show the RMSE, MAE, and R-Square line charts of various surrogate models for different sample sizes. The same color indicates the same variable-fidelity surrogate model. The solid line represents the prediction accuracy line chart of the model constructed on the nested Latin hypercube design based on translation propagation, while the dashed line of the same color shows the prediction accuracy of the model on the non-nested design. In Fig. 3.11a and b, any solid line is below the corresponding dashed line of the same color, meaning that the RMSE and MAE values of the model established on the proposed experimental design method are smaller than those on the non-nested design. In Fig. 3.11c, the

Table 3.1 Prediction accuracy of variable-fidelity surrogate models under different experimental design methods

Test. func	Sp. size	Diff.-error-criteria	DMF-LSSVR		DMF-RSM		DMF-kriging		DMF-RBF		Kri	RSM	RBF
			Nested	Non-nested	Nested	Non-nested	Nested	Non-nested	Nested	Non-nested			
	10	RMSE	4.3293	4.0721	4.2945	7.3788	5.107	5.5415	6.1271	8.2371	11.246	11.501	13.149
		MAE	11.422	12.361	13.872	17.396	10.571	13.872	13.555	20.217	19.908	20.02	26.546
		R-square	0.9558	0.9008	0.9045	0.9049	0.9061	0.7577	0.7595	0.521	0.7578	0.7469	0.6691
	20	RMSE	2.8459	2.7433	4.2092	5.8509	2.9768	3.2216	3.6902	4.8198	9.1738	9.9586	12.882
		MAE	7.8155	8.8663	8.4157	16.033	7.9979	9.1227	8.3717	12.143	20.273	20.299	25.528
		R-square	0.9807	0.982	0.873	0.7913	0.9409	0.9291	0.8883	0.8678	0.8447	0.8167	0.7035
SC	30	RMSE	1.3701	3.6458	2.6492	4.6839	2.2865	2.5619	3.3086	3.4249	1.7197	6.8865	8.1599
		MAE	3.7347	9.9467	5.6949	9.2667	8.4527	7.7845	9.2935	6.7539	3.3169	10.157	21.087
		R-square	0.9955	0.9684	0.9295	0.8632	0.9707	0.965	0.9038	0.9288	0.9943	0.9106	0.8757
	40	RMSE	1.1637	2.1986	2.8764	3.808	1.238	2.6952	2.6171	4.1389	2.9959	11.104	11.807
		MAE	3.0184	6.1814	4.2819	6.433	2.5308	8.9848	6.8797	10.101	10.274	21.657	27.392
		R-square	0.9968	0.9885	0.9108	0.8629	0.9927	0.9627	0.9604	0.9005	0.9828	0.7753	0.7494
	10	RMSE	14.763	29.759	38.991	42.017	32.684	32.624	96.779	119.31	62.489	67.453	41.829
		MAE	33.388	73.704	74.926	84.731	52.042	62.793	223.67	245.09	143.16	153.04	126.66
		R-square	0.8975	0.2291	0.356	0.3619	0.7313	−0.1408	−5.377	−5.3854	0.5835	0.5146	0.8088
	20	RMSE	8.0317	12.922	20.686	37.513	16.575	19.499	78.218	84.321	18.389	27.209	21.853
		MAE	16.241	33.178	34.307	99.425	40.325	40.748	161.87	197.19	45.044	45.177	71.479
		R-square	0.9951	0.9844	0.9522	0.3184	0.7484	0.6878	−0.7336	−2.0947	0.9622	0.9173	0.9468
MBR	30	RMSE	6.8982	7.8275	16.085	21.018	13.218	15.961	54.151	65.115	8.7226	17.787	6.3823
		MAE	12.564	18.149	27.755	41.382	33.984	48.561	144.2	239.07	27.437	28.414	15.617
		R-square	0.9936	0.9938	0.9711	0.8173	0.8871	0.9035	−0.1203	−0.7186	0.9708	0.9646	0.9954
	40	RMSE	4.5282	6.5391	15.327	21.889	11.094	11.588	40.942	49.504	8.6758	26.716	14.792
		MAX	7.9263	15.678	25.200	39.318	22.522	25.845	122.68	168.63	26.616	60.519	53.499
		R-square	0.9969	0.9916	0.8694	0.7742	0.9512	0.9461	0.6056	−0.3093	0.9916	0.9207	0.9756

(continued)

Table 3.1 (continued)

Test. func	Sp. size	Diff.-error-criteria	DMF-LSSVR		DMF-RSM		DMF-kriging		DMF-RBF		Kri	RSM	RBF
			Nested	Non-nested	Nested	Non-nested	Nested	Non-nested	Nested	Non-nested			
	10	RMSE	2.7082	0.4066	0.5698	2.7661	1.4697	2.9949	4.5638	8.986	0.7377	0.7311	24.858
		MAE	9.1397	1.2653	1.4263	9.9947	3.433	6.5838	11.787	31.751	1.6644	1.6397	59.805
		R-square	0.9905	0.9998	0.9998	0.9998	0.9996	0.9958	0.9943	−0.9792	0.9998	0.9998	0.7524
	15	RMSE	0.2563	0.5618	0.3525	1.0771	0.8481	1.9773	3.6999	4.2352	0.4892	0.4804	12.476
		MAE	0.743	1.5105	0.9067	2.8597	1.939	6.2852	12.964	12.484	0.8999	0.8988	29.982
		R-square	0.9999	0.9997	0.9999	0.9998	0.9999	0.9985	0.9972	0.9948	0.9999	0.9999	0.9382
Booth	20	RMSE	0.2261	0.2383	0.3137	0.935	0.9828	1.4107	2.4373	3.5211	0.3437	0.3488	7.8121
		MAE	0.4734	0.5612	0.5616	2.646	3.7747	3.8145	7.1245	9.1886	0.6259	0.5898	20.614
		R-square	1	0.9999	0.9999	0.9998	0.9998	0.9996	0.9985	0.9961	0.9999	0.9999	0.9753
	25	RMSE	0.1744	0.2547	0.3019	0.8721	0.8893	0.8219	1.8568	2.1108	0.3343	0.3977	6.4471

Table 3.2 Prediction accuracy of variable-fidelity surrogate models under different experimental design methods (Continued)

Test. func	Sp. size	Diff.-error-criteria	DMF-LSSVR		DMF-RSM		DMF-kriging		DMF-RBF		Kri	RSM	RBF
			Nested	Non-nested	Nested	Non-nested	Nested	Non-nested	Nested	Non-nested			
	10	RMSE	0.0871	0.1702	0.9382	1.271	0.6483	0.8746	0.7151	1.0095	0.7377	0.7311	24.858
		MAE	0.1976	0.4004	2.2414	3.0419	1.4336	1.587	2.0207	2.7903	1.6644	1.6397	59.805
		R-square	1	1	0.8379	0.7748	0.9274	0.8764	0.9378	0.7892	0.9998	0.9998	0.7524
Bohachevsky	20	RMSE	0.0624	0.1783	0.5341	0.976	0.5918	0.7818	0.4738	0.6702	0.4168	0.4052	25.311
		MAE	0.2265	0.5489	1.1251	3.3837	1.0204	2.3179	1.0073	1.1498	0.7857	0.6766	81.1737
		R-square	1	0.9999	0.966	0.8762	0.9455	0.9255	0.9685	0.9391	0.9999	0.9999	0.8961
	30	RMSE	0.0478	0.0527	0.5365	0.6581	0.5606	0.7146	0.3984	0.6539	0.3582	0.3677	9.6732
		MAE	0.1258	0.1025	1.0938	2.0728	1.6283	1.4249	0.7824	12.576	0.7407	0.6655	29.119
		R-square	1	1	0.9661	0.9492	0.9608	0.9357	0.9846	0.9382	0.9999	0.9999	0.9848
	40	RMSE	0.0158	0.0475	0.3661	0.6173	0.5214	0.6355	0.3928	0.5979	0.3138	0.3625	9.4182
		MAE	0.0343	0.0941	0.7419	1.1339	12.288	1.6499	0.9434	1.3446	0.7520	0.6989	25.551
		R-square	1	1	0.9751	0.9544	0.9625	0.9301	0.9835	0.9537	0.9999	0.9999	0.9856
HM	10	RMSE	4.556	7.7095	18.271	21.626	11.156	18.832	7.1841	8.2947	26.268	26.292	27.904
		MAE	10.533	16.062	53.847	68.33	29.938	61.964	12.165	15.565	83.874	83.925	97.253
		R-square	0.9915	0.9812	0.8684	0.81559	0.9381	0.8506	0.9819	0.9751	0.7289	0.7281	0.7017
	20	RMSE	2.9168	4.3904	9.5351	14.582	5.8683	10.036	5.5916	6.8651	17.585	17.958	8.0555
		MAE	8.8167	14.538	26.305	48.238	17.484	35.055	16.863	16.559	43.934	43.549	17.683
		R-square	0.9968	0.9924	0.9553	0.9288	0.9907	0.9626	0.9894	0.9843	0.8775	0.8723	0.9743
	30	RMSE	0.9061	2.0056	6.9475	8.7652	2.183	6.9462	3.5283	4.0485	10.614	20.195	14.772
		MAE	2.2973	4.5288	14.584	19.285	7.5718	23.931	11.045	7.8119	18.379	59.414	34.777
		R-square	0.9996	0.9984	0.9809	0.9612	0.9979	0.9832	0.9951	0.9947	0.9158	0.8413	0.9135
	40	RMSE	0.5865	1.0416	5.9089	7.7461	1.1659	3.9186	2.1092	3.4517	2.434	18.847	6.8262
		MAE	1.1973	2.6794	12.823	16.767	4.3421	12.658	5.3952	10.935	1.6622	51.471	21.103
		R-square	0.9999	0.9995	0.9779	0.9762	0.9994	0.9943	0.9975	0.9946	0.9099	0.8594	0.9815
	10	RMSE	5.4222	9.1883	19.270	29.959	9.2148	26.941	8.2112	21.699	144.37	114.77	90.772
		MAE	13.014	21.626	33.606	74.266	25.616	100.04	19.214	53.697	407.84	283.04	303.19
		R-square	0.9944	0.9910	0.9134	0.9081	0.9926	0.9300	0.9897	0.9559	0.0015	0.2557	0.2969

(continued)

Table 3.2 (continued)

Test. func	Sp. size	Diff.-error-criteria	DMF-LSSVR		DMF-RSM		DMF-kriging		DMF-RBF		Kri	RSM	RBF
			Nested	Non-nested	Nested	Non-nested	Nested	Non-nested	Nested	Non-nested			
3d Rosenbrock (n=3)	20	RMSE	3.9476	7.6795	11.795	21.609	6.5262	23.926	7.1768	19.439	88.289	96.161	−68.217
		MAE	10.693	20.097	30.434	80.139	22.901	89.051	23.394	70.256	268.46	235.01	215.12
		R-square	0.9985	0.9956	0.9915	0.9835	0.9964	0.9696	0.9977	0.9723	0.5691	0.4528	0.5930
	30	RMSE	1.8620	6.0555	7.2084	15.651	4.3886	14.955	5.7100	16.445	48.118	77.083	32.698
		MAE	4.7013	18.719	24.929	29.638	16.149	54.839	20.521	41.904	99.701	216.25	80.427
		R-square	0.9998	0.9952	0.9941	0.9723	0.9981	0.9806	0.9979	0.9736	0.6628	0.4712	0.8514
	40	RMSE	0.8964	4.4276	5.2390	10.689	3.3411	7.4140	5.0509	11.267	25.279	48.413	24.607
		MAE	2.0593	10.551	17.717	22.737	12.139	16.758	18.633	37.421	90.189	100.23	50.161
		R-square	0.9999	0.9970	0.9981	0.9856	0.9989	0.9936	0.9978	0.9850	0.8708	0.5406	0.8536
	10	RMSE	2.9013	5.4683	53.106	73.484	11.566	19.502	29.741	35.751	36.213	114.41	64.326
		MAE	9.2265	15.726	97.206	143.99	20.93	43.496	62.451	71.437	103.64	212.04	152.56
		R-square	0.9971	0.9898	0.0388	0.8837	0.9192	0.7835	0.6883	0.5533	0.5525	3.3888	0.3592
Borehole model (n=8)	20	RMSE	2.7885	4.6383	55.219	74.003	7.068	11.785	23.932	27.295	26.134	189.45	59.978
		MAE	8.7586	15.226	136.04	210.49	16.093	36.001	58.002	53.835	89.019	575.95	105.45
		R-square	0.9973	0.9926	−0.3552	−0.6896	0.9632	0.9306	0.7405	0.6296	0.7668	8.8264	0.2191
	50	RMSE	1.3915	3.2621	10.221	11.989	5.2101	6.9062	17.808	19.152	13.477	14.833	60.972
		MAE	5.1683	12.142	19.824	33.397	12.425	20.485	30.355	52.457	28.077	33.181	133.29
		R-square	1	0.9964	0.9418	0.9267	0.9895	0.973	0.7889	0.6923	0.9378	0.9246	−0.2739

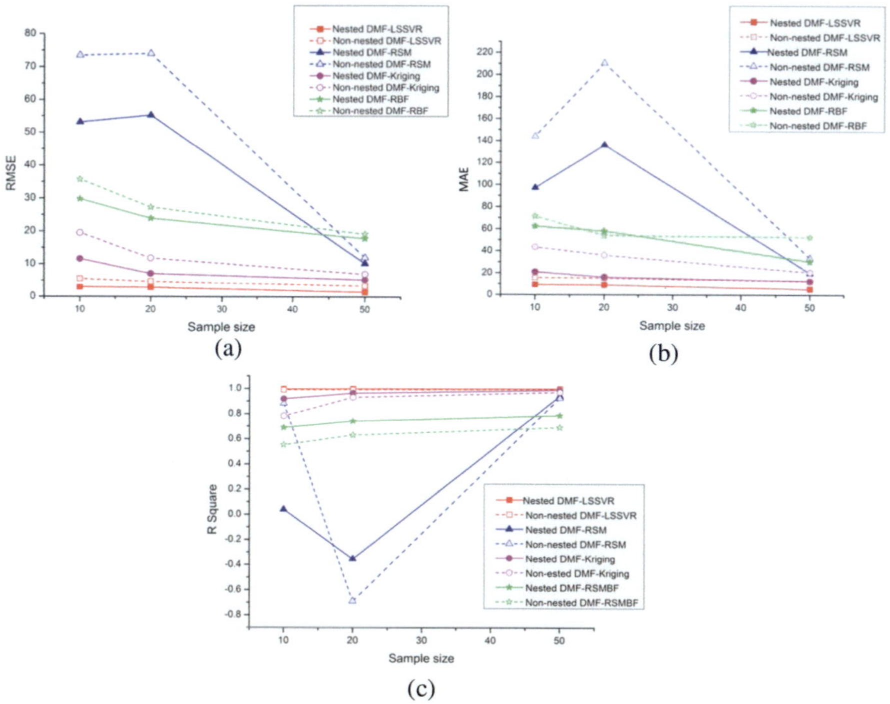

Fig. 3.11 Line charts of RMSE, MAE, and R-Square for variable-fidelity surrogate models of the deep hole model

solid line of the same color is above the dashed line, indicating that the R-Square value of the model established on the proposed experimental design method is larger than that on the non-nested design. Smaller RMSE and MAE values and a larger R-Square value indicate higher precision, demonstrating that the nested Latin hypercube design proposed in this chapter is effective in improving the prediction accuracy of variable-fidelity surrogate models.

3.3 Engineering Test Case

This section employs the NACA 0012 airfoil dynamics modeling case to validate the proposed method. Specifically, a least squares output mapping model based on a nested Latin hypercube design is used to construct the lift coefficient of the NACA 0012 airfoil. The design variables are Mach number Ma and angle of attack α, with the range of $0.3 \le Ma \le 0.5$ and $0^\circ \le \alpha \le 10^\circ$. High- and low-fidelity analysis models (both using Fluent 14.5 CFD simulations) are employed for sample acquisition. The LF analysis model has approximately 15,000 grid points, as shown in Fig. 3.12a, while the corresponding HF analysis model has approximately 74,000 grid points, as shown in Fig. 3.12b.

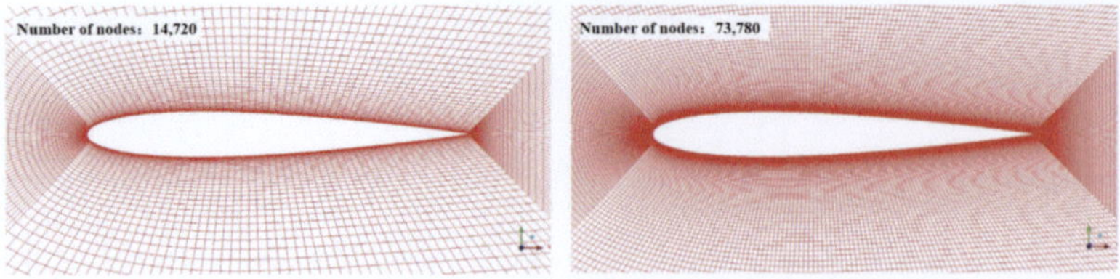

Fig. 3.12 High- and low-fidelity CFD analysis models for NACA 0012 airfoil **a** LF analysis model **b** HF analysis model

Table 3.3 Non-nested DOE LF sample data

No.	α(deg.)	Ma	C_{LL}	No.	α(deg.)	Ma	C_{LL}
1	0.34	0.40	0.0179	16	9.66	0.42	0.4452
2	3.10	0.36	0.1243	17	3.79	0.32	0.1176
3	2.76	0.47	0.1975	18	7.24	0.44	0.4080
4	5.17	0.35	0.1892	19	6.21	0.38	0.2641
5	2.41	0.40	0.1220	20	10.00	0.37	0.3508
6	6.55	0.48	0.4547	21	2.07	0.50	0.1715
7	1.72	0.31	0.0511	22	3.45	0.43	0.2013
8	7.59	0.41	0.3642	23	0.69	0.47	0.0506
9	1.03	0.36	0.0423	24	9.31	0.45	0.5021
10	8.28	0.38	0.3312	25	4.48	0.39	0.2087
11	5.52	0.42	0.2951	26	6.90	0.34	0.2286
12	5.86	0.31	0.1643	27	8.62	0.49	0.5734
13	0.00	0.33	0.0002	28	8.97	0.33	0.2616
14	4.83	0.46	0.3190	29	7.93	0.30	0.1973
15	4.14	0.49	0.3185	30	1.38	0.44	0.0866

The experimental sample size for constructing the variable-fidelity surrogate model is 30 LF sample points and 15 HF sample points. The non-nested design sample points and their corresponding high- and low-fidelity responses are shown in Tables 3.3 and 3.4, where "C_{LL}" represents the LF response and "C_{LH}" represents the HF response. The sample point distribution for the nested design based on translation propagation is shown in Fig. 3.13, and the corresponding HF responses are listed in Table 3.5. The validation samples in Table 3.6 are used to calculate the precision metrics RRMSE and RMAE for the models constructed using different methods.

Table 3.7 shows that the least squares support vector regression output mapping model under a nested design has the smallest RRMSE and RMAE values. Both numerical and engineering test cases indicate that the proposed nested design significantly contributes to improving the accuracy of variable-fidelity surrogate models.

Table 3.4 Non-nested DOE HF sample data

No.	α(deg.)	Ma	C_{LH}	No.	α(deg.)	Ma	C_{LH}
1	7.14	0.5	0.5635	9	6.74	0.44	0.1066
2	8.57	0.36	0.3216	10	10	0.3	0.2375
3	10.00	0.44	0.5288	11	0.05	0.5	0.0047
4	1.43	0.33	0.0494	12	3.96	0.34	0.1433
5	2.86	0.47	0.2109	13	2.33	0.42	0.1344
6	0.00	0.41	0.0002	14	6.21	0.36	0.2461
7	4.29	0.39	0.2073	15	10	0.5	0.6603
8	5.71	0.30	0.1560				

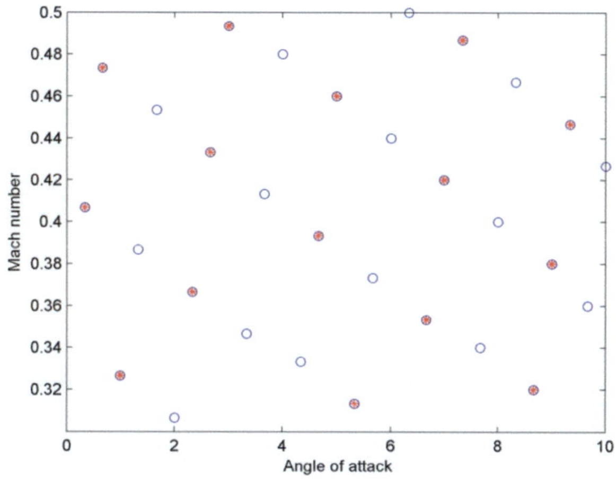

Fig. 3.13 Sample point distribution of nested design

3.4 Summary

This chapter proposes an experimental design method for variable-fidelity surrogate models, namely, nested Latin hypercube design based on translation propagation. The basic idea is to first construct a simple seed design with nested relationships and then translate and propagate it across the entire design space. Additionally, strategies for removing excess points are developed to achieve nested Latin hypercube designs of any sample size. Numerical and engineering test cases, including the NACA 0012 airfoil dynamics modeling, validate the proposed method. The results confirm the rationality and effectiveness of the proposed method, demonstrating that nested Latin hypercube design can rapidly and accurately determine the distribution of high- and low-fidelity samples, providing support for constructing high-quality variable-fidelity surrogate models.

Table 3.5 Nested design sample data

No.	α(deg.)	Ma	C_{LL}	C_{LH}	No.	α(deg.)	Ma	C_{LL}	C_{LH}
1	2.00	0.31	0.0221		16	0.33	0.41	0.0129	0.0055
2	5.33	0.31	0.0381	0.0562	17	3.67	0.41	0.1126	
3	8.67	0.32	0.0764	0.1002	18	7.00	0.42	0.2386	0.1926
4	1.00	0.33	0.0141	0.0126	19	19	10.00	0.43	0.3509
5	4.33	0.33	0.0318		20	2.67	0.43	0.0929	0.0463
6	7.67	0.34	0.2449		21	6.00	0.44	0.1864	
7	3.33	0.35	0.6347		22	9.33	0.45	0.0388	
8	6.67	0.35	0.0223	0.0325	23	1.67	0.45	0.0388	
9	9.67	0.36	0.2866		24	5.00	0.46	0.1724	0.1441
10	2.33	0.37	0.2080	0.1826	25	8.33	0.47	0.3056	
11	5.67	0.37	0.1134		26	0.67	0.47	0.0224	
12	9.00	0.38	0.1189	0.1213	27	4.00	0.48	0.1421	
13	1.33	0.39	0.0259		28	7.33	0.49	0.2897	0.2288
14	4.67	0.39	0.1051	0.0918	29	3.00	0.49	0.1198	0.1296
15	8.00	0.40	0.2409		30	6.33	0.50	0.2818	

Table 3.6 Validation sample and response data

No.	α(deg.)	Ma	C_L	No.	α(deg.)	Ma	C_L
1	0	0.39	0.0002	9	3.58	0.43	0.2156
2	0.71	0.46	0.0057	10	8.57	0.30	0.2214
3	1.43	0.33	0.0494	11	9.29	0.36	0.3415
4	2.14	0.43	0.1464	12	10	0.49	0.6374
5	2.86	0.34	0.1044	13	3.57	0.50	0.3015
6	0	0.5	0.0004	14	4.29	0.41	0.2312
7	8.33	0.31	0.2315	15	5.00	0.37	0.2143
8	4.77	0.5	0.3970				

Table 3.7 Comparison of NACA 0012 Airfoil Modeling Structures

Accuracy metrics	LSSVR-OM (nested)	LSSVR-OM (non-nested)	AL-VFM [5]	MMD-VFM [5]	Kriging
RMAE	0.183	0.256	0.366	0.406	0.387
RRMSE	0.096	0.141	0.139	0.162	0.222

References

1. Luk-Cyr J, El-Bawab R, Champliaud H et al (2017) Sequential approximate optimization of industrial hammer peening using finite element simulations. Struct Multidiscipl Opt 55:767–778
2. Gano SE, Renaud JE, Martin JD et al (2006) Update strategies for kriging models used in variable fidelity optimization. Struct Multidiscipl Opt 32:287–298
3. Zhang ZGD, Wang M (2011) A new and better metamodel utilizing difference between high-and low-fidelity model. J Northwestern Polytech Univ 29(2):176–182
4. Gong C, Yuan J, Gu L et al (2006) Improving variable complexity model for aerodynamic analysis. J Northwestern Polytech Univ 24(4):532–535
5. Zhou Q, Shao X, Jiang P et al (2016) An active learning metamodeling approach by sequentially exploiting difference information from variable-fidelity models. Adv Eng Inf 30(3):283–297

Chapter 4
Nested Maximin Designs Based on Successive Local Enumeration and Discrete Optimization

Computer simulation, leveraging physical behavior models and scientific computations, is a cornerstone of modern mechanical product design [1]. It reduces the need for costly physical experiments, accelerates development, and provides critical data for design optimization. However, as finite element models grow in complexity—with finer mesh elements and larger numbers of elements—computational costs increase exponentially, posing challenges for design optimization [2]. Variable-fidelity surrogate modeling addresses this by combining low-fidelity (LF) models, which are computationally efficient, with high-fidelity (HF) models, which ensure accuracy, to construct cost-effective surrogate models.

Experimental design is pivotal in variable-fidelity modeling, as the selection of sample points directly influences the accuracy of surrogate models. Traditional single-fidelity methods, such as Uniform Design [3], Orthogonal Array Sampling [4], and Latin Hypercube Design (LHD) [5], are not directly applicable to variable-fidelity contexts. This chapter proposes a nested LHD method, termed SLE-NGHS, which integrates Successive Local Enumeration (SLE) for LF sampling and an improved Novel Global Harmony Search (NGHS) algorithm for HF sampling. The method is evaluated for space-filling properties, computational cost, and prediction accuracy, and its effectiveness is demonstrated through a natural gas pressure vessel optimization case study.

4.1 Nested Latin Hypercube Experimental Design Method with SLE and NGHS

In this section, a nested Latin hypercube design method (SLE-NGHS) is proposed. This method uses the continuous local enumeration method to generate low-fidelity (LF) sampling points, and then applies an improved NGHS algorithm to select several appropriate sampling points from the LF sampling points as high-fidelity (HF) sampling points. First, the continuous local enumeration method for constructing LF

J. Yi et al., *Variable-Fidelity Surrogate*, Engineering Applications of Computational Methods 25, https://doi.org/10.1007/978-981-95-5527-7_4

sampling points is introduced. Then, the problem of selecting HF sampling points is modeled to obtain a combinatorial optimization model. Finally, the improved NGHS algorithm is applied to solve this model.

4.1.1 Constructing LF Sampling Points Using Continuous Local Enumeration Method

The Successive Local Enumeration (SLE) method, initially proposed by Zhu et al. [6], is an effective Latin hypercube design (LHD) sampling method. It can generate a uniformly distributed maximin Latin hypercube sample set, which maximizes the minimum distance between sample points.

The following introduces the SLE method with the help of a two-dimensional design space: Imagine the two-dimensional design space as a chessboard. Maximin Latin hypercube sampling (Maximin LHD) can be analogized to a chess problem—placing n chess pieces on an $n \times n$-grid chessboard such that the pieces do not overlap, and the minimum Euclidean distance d between all pieces is maximized. The coordinate positions of the pieces on the chessboard are denoted as $(x_k, y_k) \in \{1, 2, \ldots, n\}^2, k = 1, 2, \ldots, n$.

The workflow of the SLE method is as follows:

1. For the first column, the positions of sample points can be randomly selected. Suppose the coordinates of the first sample point P_1 are $(i_1, 1)$, where $i_1 \in \{1, 2, \ldots, n\}$, and it is added to the sample point set $P = \{P_1\}$.
2. For the second column, since the position in row i_1 has already been occupied by P_1, one of the remaining rows should be selected as the position for the second sample point. By calculating the distances between the first sample point and all points in the remaining rows of the second column, it is found that the point at $(i_2, 2)$ has the farthest distance from P_1. Therefore, the position of the second sample point P_2 is chosen as $(i_2, 2)$, which is then added to the sample point set $P = \{P_1, P_2\}$.
3. For the third column, let $s = \{1, 2, \ldots, n\}$. Currently, the sample set is $P = \{P_1, P_2, \ldots, P_n\}$, so the s-th sample point should be placed in the remaining $(m - s + 1)$ positions. The characteristic value of a row in the s-th column is defined as the minimum distance from that row's point to all previously selected sample points. The sample point for the s-th column is then the point in the row with the maximum characteristic value, denoted as P_s, which is added to the sample set to form $P = \{P_1, P_2, \ldots, P_{s-1}, P_s\}$.
4. Repeat Step 3 until the $n - 1$-th sample point is added to the sample set.
5. Select the n-th sample point. Since only one row remains available at this point, the last sample point is chosen as the remaining row in the n-th column.

To describe the workflow of the successive local enumeration method in more detail, we take a case with a two-dimensional design space and the number of sample points $n = 5$ as an example for further illustration, as shown in Fig. 4.1.

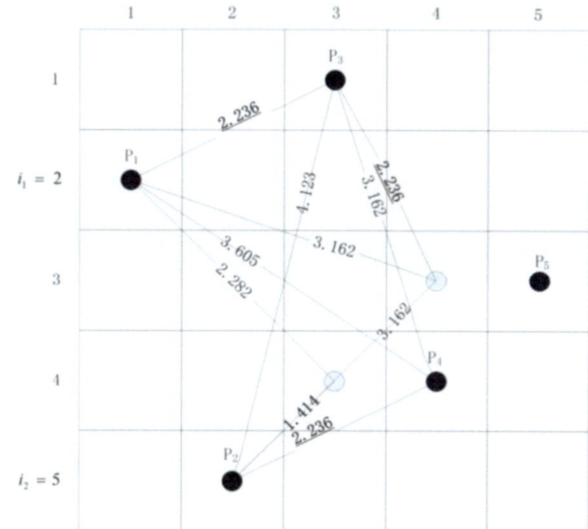

Fig. 4.1 Selection of 5 sampling points in a two-dimensional space

Step 1: Randomly select the point with row number $i = 2$ as the first sample point P_1. Step 2: In the second column, the available rows are $\{1, 3, 4, 5\}$. Calculate the distances between the points corresponding to these rows and the first sample point P_1. Select the point with the maximum distance, where the row number is $i = 5$, as the second sample point P_2. Step 3: For the third column, the remaining available rows are $\{1, 3, 4\}$. Calculate the distances between the point $(1, 3)$ and the existing points $\{P_1, P_2\}$: $d((1, 3), P_1) = 2.236 < d((1, 3), P_2) = 4.123$. Thus, the characteristic value for row 1 is 2.236. Similarly, the characteristic values for rows 3 and 4 are calculated as 2 and 1.414, respectively. By comparison, row 1 has the maximum characteristic value. Therefore, the sampling point for this column is selected as row 1, i.e., $P_3 = (1, 3)$. Step 4: Similar to the third column, through comparison, it is found that the characteristic value of row 4 is the largest. Therefore, the point in row 4 of the fourth column is selected as the sampling point, i.e., $P_4 = (4, 4)$. Step 5: Since only the last row remains available, the third row is selected as the final sampling point in the fifth column, i.e., $P_5 = (3, 5)$.

4.1.2 Modeling the HF Sampling Point Selection Problem

The HF sampling point selection problem is to select n_2 HF sampling points from n_1 LF sampling points such that the minimum distance between HF sampling points is maximized. Assuming S is the set of indices of the sampling points selected from LF, the decision variables for this problem are:

$$S = (s_1, s_2, \ldots, s_{n_2}), s_i \in [1, n_1], i \in [1, n_2]. \tag{4.1}$$

Therefore, the optimization model for this problem can be expressed as

$$\begin{aligned} &\text{Max:}\quad f(S) = \min(d(x_{s_i}, x_{s_j})) \\ &\text{s.t.}\quad s_i \neq s_j, i, j \in [1, n_2] \quad \text{and} \quad i \neq j. \end{aligned} \tag{4.2}$$

For a nested design with size (n_1, n_2), its solution space size is $C_{n_1}^{n_2}$. For example, a nested design of (100, 10) has $1.73e + 13$ feasible solutions, making it a combinatorial optimization problem that is extremely difficult to solve. This problem shares the same solution space size as the NP-hard Maximum Diversity Problem [7]. To solve this problem efficiently, this chapter proposes an improved NGHS algorithm.

4.1.3 Solving the HF Sampling Point Selection Problem Using the Improved NGHS Algorithm

The NGHS algorithm, an improved HS algorithm proposed by Zou et al., has its detailed steps introduced in Chap. 2 and will not be repeated here. Since the NGHS algorithm is designed for continuous optimization problems, while the decision variables in this problem are the indices of sampling points (belonging to integers), the NGHS algorithm cannot be directly applied. Here, a straightforward approach is adopted to convert real numbers into the nearest integers [8].

To satisfy the constraint $s_i \neq s_j$, a Check and Repair operation is performed after generating new harmonies. In this operation, once a new harmony is generated, it is first checked for compliance with the constraint conditions. If not satisfied, redundant variables are reassigned until the constraints are met. The pseudocode for the Check and Repair operation is shown in Fig. 4.2.

Algorithm 6 Inspection and repair operations

```
Start
B = unique(harmony) /* The function returns all non-repeating elements to the set B */
if length(B) ≠ n2 /* n2 is the number of decision variables */ then
    for i = 1 : length(B) do
        [K] = find(harmony = B(i)) /* Find the indices of all redundant elements */
        for j = 2 : size(K) do
            flag = FALSE
            while flag = FALSE do
                harmony(K(j))]) = randi(1, n) /* Replace redundant elements with randomly generated integers */
                C = intersect({harmony(K(j))}, B) /* The set C is the intersection of two sets */
                flag = isempty(C) /* If C is empty, the isempty function returns TRUE, otherwise FALSE */
                if flag = FALSE then
                    B = unique(harmony) /* Update set B */
                end if
            end while
        end for
    end for
end if
End
```

Fig. 4.2 Pseudocode for check and repair operation

4.1.4 Solving the HF Sampling Point Selection Problem Using the Improved NGHS Algorithm

The nested Latin hypercube design method based on the successive local enumeration method and the improved NGHS algorithm has been described in detail above. To summarize the main idea of this method, we present the overall workflow diagram of this experimental design method, as shown in Fig. 4.3.

4.2 Numerical Experimental Results

To validate the effectiveness of the proposed SLE-NGHS nested experimental design method, this section conducts a multi-faceted comparison between the SLE-NGHS method and several other nested experimental design methods. The methods involved in the comparison include the Monte Carlo method, the nested TPLHD method [9], and the ESE-GroupRand method [10]. The evaluation criteria mainly cover three

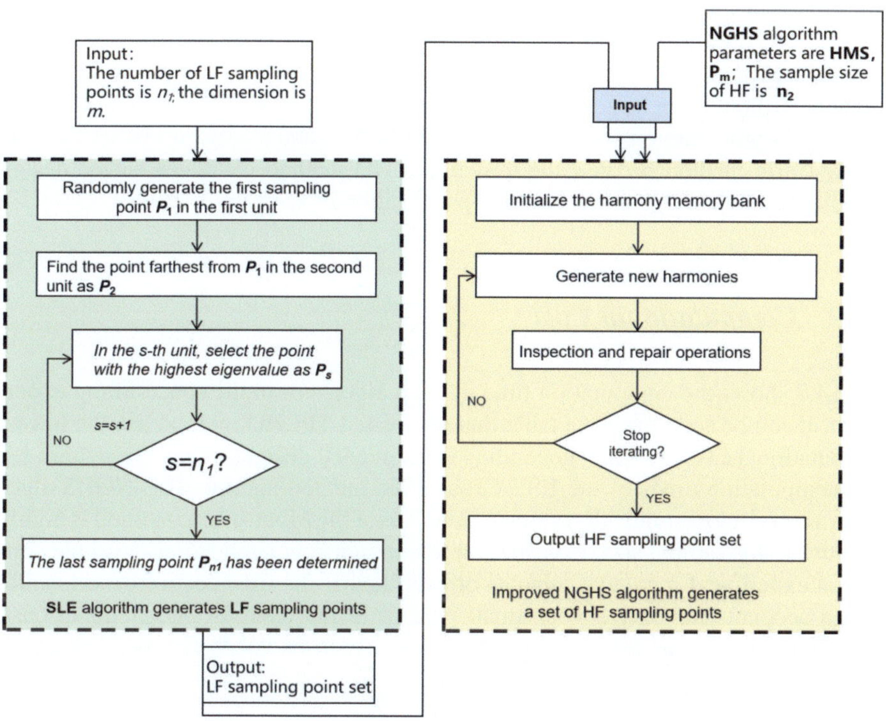

Fig. 4.3 Overall flowchart of the algorithm

aspects: space-filling performance, computational cost, and prediction accuracy. A comparative analysis of these three aspects is provided below.

4.2.1 Space-Filling Property

As previously introduced, the space-filling property of nested designs can be objectively evaluated by the scaled maximin distance. A larger maximin distance indicates better space-filling performance of the experimental design.

In this experiment, the SLE-NGHS algorithm, ESE-GroupRand algorithm, and nested TPLHD algorithm were each run 10 times, and their optimal values, average values, and worst values were obtained. Since both the SLE-NGHS and ESE-GroupRand algorithms involve optimization processes, the number of iterations for both algorithms was set to 10,000 for fairness. As the Monte Carlo method is a stochastic approach that requires a large number of repeated trials to achieve results close to the true level, this method was repeated 50,000 times in this experiment, with its optimal values, average values, and worst values recorded.

Table 4.1 presents the scaled maximin distances of all methods under different dimensions and sample sizes. It can be seen from the table that the results of the SLE-NGHS algorithm are superior to those of other methods, followed by the ESE-GroupRand method, and then the nested TPLHD method and the Monte Carlo method. Although the nested TPLHD method and the Monte Carlo method do not involve an optimization process, the nested TPLHD method performs better than the Monte Carlo method because the nested TPLHD method generates nested design schemes through uniform translational propagation of seeds.

4.2.2 Computational Cost

Table 4.2 shows the computation time of each algorithm in the space-filling experiment. It can be seen from the table that the nested TPLHD method has the lowest computational cost, basically generating experimental designs of any size within 1 s. The computation times of the ESE-GroupRand method and the SLE-NGHS algorithm are very close, while the computation time of the Monte Carlo method is highly sensitive to the sample size. Considering that in practical applications, a single simulation experiment may take minutes or even hours, the time cost of experimental design accounts for only a very small portion of the overall computational cost. Therefore, the computational costs of the above experimental design methods are all acceptable.

Table 4.1 Comparison of maximin distances of various algorithms

Dimension	Size	Monte Carlo			Nested TPLHD			ESE-GroupRand			SLE-NGHS		
	(n_2, n_z)	Best	Mean	Worst	Best	Mean	Worst	Best	Mean	Worst	Best	Mean	Worst
2	100,10	0.2737	0.2535	0.2298	0.4450	0.3340	0.3015	0.6231	0.6100	0.5924	0.7240	0.6987	0.6798
	100,20	0.2796	0.2451	0.2244	0.4455	0.2993	0.1572	0.6147	0.6061	0.5828	0.7317	0.6960	0.6711
	100,30	0.2690	0.2465	0.2084	0.4975	0.2923	0.1942	0.6276	0.6133	0.5818	0.7252	0.6945	0.6595
	100,40	0.2974	0.2508	0.2227	0.4756	0.2961	0.2252	0.5949	0.5827	0.5645	0.7038	0.6449	0.6032
3	100,10	0.3527	0.3281	0.2968	0.5254	0.3659	0.2326	0.7317	0.7206	0.7015	0.7811	0.7663	0.7501
	100,20	0.3841	0.3384	0.3199	0.5110	0.3726	0.2151	0.7269	0.7093	0.6874	0.7799	0.7629	0.7383
	100,30	0.3624	0.3357	0.2955	0.5355	0.3759	0.2129	0.7088	0.6922	0.6741	0.7833	0.7582	0.7367
	100,40	0.3863	0.3191	0.2850	0.5221	0.3831	0.2350	0.7024	0.6846	0.6617	0.7592	0.7336	0.6965
4	100,10	0.4353	0.4140	0.3988	0.6257	0.4661	0.2362	0.8445	0.8202	0.7924	0.8467	0.8254	0.8154
	100,20	0.4459	0.4136	0.3825	0.6197	0.4651	0.2230	0.8247	0.8022	0.7715	0.8581	0.8305	0.8154
	100,30	0.4873	0.4277	0.3994	0.6480	0.4481	0.2804	0.8199	0.7969	0.7639	0.8480	0.8348	0.8118
	100,40	0.4693	0.4174	0.3875	0.6488	0.4544	0.2249	0.8253	0.7942	0.7639	0.8486	0.8158	0.7994
6	100,10	0.6059	0.5887	0.5699	0.8002	0.5898	0.3860	0.9791	0.9658	0.9503	0.9886	0.9729	0.9624
	100,20	0.6275	0.5717	0.5461	0.7620	0.5883	0.2643	0.9601	0.9485	0.9307	0.9764	0.9644	0.9444
	100,30	0.5920	0.5785	0.5723	0.7580	0.5642	0.3003	0.9641	0.9437	0.9324	0.9926	0.9684	0.9385
	100,40	0.5959	0.5774	0.5666	0.7133	0.5661	0.3212	0.9715	0.9452	0.9304	0.9741	0.9581	0.9514

Table 4.2 Comparison of computational costs of various algorithms (Unit: Seconds)

n_1	n_2	Method	$m = 2$	$m = 3$	$m = 4$	$m = 6$
100	10	Monte Carlo	37.661	47.452	48.880	48.726
		ESE-GroupRand	84.921	85.560	86.158	85.194
		Nested TPLHD	0.054	0.058	0.054	0.123
		SLE-NGHS	83.669	84.193	83.622	83.661
100	20	Monte Carlo	88.180	104.749	105.148	106.115
		ESE-GroupRand	84.500	88.516	88.955	89.272
		Nested TPLHD	0.053	0.057	0.053	0.119
		SLE-NGHS	86.441	85.866	86.129	87.565
100	30	Monte Carlo	213.226	187.071	188.952	189.056
		ESE-GroupRand	93.526	93.692	93.575	93.896
		Nested TPLHD	0.051	0.062	0.057	0.123
		SLE-NGHS	90.527	89.936	89.982	87.790
100	40	Monte Carlo	383.819	298.058	299.910	301.493
		ESE-GroupRand	99.750	100.153	99.950	99.151
		Nested TPLHD	0.055	0.063	0.058	0.124
		SLE-NGHS	98.615	97.383	97.708	97.509

4.2.3 Prediction Accuracy

To compare the impact of different experimental design methods on the prediction accuracy of variable-fidelity surrogate models, this subsection will test several multi-peak functions F_1–F_4 from low to high dimensions. Since these functions are multi-peak with multiple minima, their landscapes are complex, making it extremely challenging to construct accurate surrogate models. Therefore, the quality of sampling has a significant impact on variable-fidelity surrogate models. In all functions, y_h represents the high-fidelity model, and y_l represents the low-fidelity model. For each test function, different nested experimental design methods are used to generate samples of the same size. To ensure fairness in the comparison, the same variable-fidelity surrogate model is employed for prediction, and the prediction accuracy is ultimately obtained. Since all variable-fidelity surrogate models are identical, the prediction accuracy can reflect which nested experimental design method is superior. The test functions used in this experiment are as follows:

1. Six-hump camel-back Function (F_1):

$$\begin{gathered} f(x_1, x_2) = 4x_1^2 - 2.1x_1^4 + \tfrac{x_1^6}{3} + x_1x_2 - 4x_2^2 + 4x_2^4 \\ y_h = f(x_1, x_2) \\ y_l = f(0.7x_1, 0.7x_2) + x_1x_2 - 65 \\ x_{1,2} \in [-2, 2]. \end{gathered} \tag{4.3}$$

2. 3D Rosenbrock Function (F_2):

$$\begin{gathered} f(x_1, x_2, x_3) = \sum_{i=1}^{2} \left[100(x_i^2 - x_{i+1}^2) + (x_i - 1)^2\right]; \\ y_h = f(x_1, x_2, x_3); \\ y_l = 90(x_1^2 - x_2^2) + 1.1(x_1 - 1)^2 + 100(x_2^2 - x_3^2) + (x_2 - 1)^2; \\ x_{1,2,3} \in [-2.048, 2.048]. \end{gathered} \tag{4.4}$$

3. Colville-Himmelblau Function (F_3):

$$\begin{gathered} f(x_1, x_2, x_3, x_4) = 100(x_1^2 - x_2)^2 + (x_1 - 1)^2 + 90(x_3^2 - x_4)^2 + (x_3 - 1)^2 \\ + 10.1(x_2 - 1)^2 + (x_4 - 1)^2 + 19.8(x_2 - 1)(x_4 - 1); \\ y_h = f(x_1, x_2, x_3, x_4); \\ y_l = 100(x_1^2 - x_2)^2 + 1.1(x_1 - 1)^2 + 90(x_3^2 - x_4)^2 + 1.1(x_3 - 1)^2 \\ +10(x_2 - 1)^2 + (x_4 - 1)^2 + 20(x_2 - 1)(x_4 - 1); \\ x_i \in [-10, 10], \quad i = 1, 2, 3, 4. \end{gathered}$$

4. Hessen Function (F_4):

$$\begin{gathered} f(x_1, \ldots, x_6) = 25(2 - x_2)^2 + (x_2 - 2)^2 + (x_3 - 2)^2 \\ + 90(4 - x_4)^2 + (x_5 - 1)^2 + (x_6 - 4)^2; \\ y_h = f(x_1, \ldots, x_6); \\ y_l = 0.6 f(x_1, \ldots, x_6) + 0.25(x_2 - 2)^2 + 0.75(x_4 - 4)^2; \\ x_i \in [0, 10], \quad i = 1, 2, \ldots, 6. \end{gathered}$$

To ensure the fairness of the comparison, each nested experimental design method was used to perform 10 independent samplings for each function, and the average values of three indicators were obtained: root mean square error (RMSE), maximum absolute error (MAE), and R^2.

Table 4.3 presents the prediction accuracy indicators of various algorithms. It can be seen from the table that the RMSE and MAE values of the SLE-NGHS algorithm are smaller than those of other methods, and the R2 value of the SLE-NGHS algorithm is closer to 1 compared with other methods. Therefore, the nested sampling method based on the SLE-NGHS algorithm achieves higher prediction accuracy than the other sampling methods.

Figure 4.4 shows a comparison between the true function response surface of the first test function, the low-fidelity surrogate model surface, the high-fidelity surrogate model surface, and the variable-fidelity surrogate model surface. It can be seen from the figure that the variable-fidelity surrogate model surface fits the true surface very closely, indicating that the variable-fidelity surrogate model can achieve a high-precision approximation of the true response surface. The high accuracy of the variable-fidelity surrogate model also benefits, to a certain extent, from the good space-filling property of the sampling points generated by the SLE-NGHS algorithm.

Table 4.3 Comparison of prediction accuracy of various algorithms

Function	Indicator	Method			
		Monte Carlo	ESE-GroupRand	Nested TPLHD	SLE-NGHS
F_1	MAE	1.913919	2.192069	5.0703283	0.647239
	RMSE	1.031082	1.005222	2.464174	0.280399
	R^2	0.982074	0.984135	0.901818	0.999725
F_2	MAE	16.66201	14.41094	18.46217	0.974460
	RMSE	8.559764	8.895734	10.15581	0.392396
	R^2	0.999600	0.999562	0.999426	0.999999
F_3	MAE	61.30853	78.81615	26.07916	0.898723
	RMSE	30.81895	32.53247	15.67686	0.393925
	R^2	0.998807	0.998646	0.999702	0.999998
F_4	MAE	124.0769	126.8029	66.12327	42.04787
	RMSE	54.78975	53.97068	34.85196	16.62686
	R^2	0.997420	0.998297	0.999237	0.999961

4.3 Engineering Case Study: Design Optimization of Pressure Vessel

In this subsection, the effectiveness of the SLE-NGHS algorithm is verified through the experimental design and modeling part in an engineering example of natural gas pressure vessel optimization design. This engineering example was first proposed by Zheng et al. [11], aiming to design a cylindrical pressure vessel for natural gas storage. Figure 4.5 shows the schematic diagram and cross-sectional view of the pressure vessel. The outer diameter of the pressure vessel is 406 mm, the main body length is 6000 mm, and the average gas pressure is 23 MPa. The elastic modulus and Poisson's ratio of the material are $E = 207$ GPa and $\mu = 0.3$, and the allowable tensile strength of the material is $\sigma_s = 414$ MPa. To enhance the strength of the pressure vessel end, the thickness from the end to the main body decreases gradually. The design goal of the pressure vessel is to minimize the manufacturing materials while ensuring that the internal volume V of the pressure vessel is not less than $0.63\ \text{m}^3$.

The design variables of this problem and their value ranges are shown in Table 4.4, and its optimization model is as follows:

$$
\begin{gathered}
\min : M \\
\text{s.t.} : \sigma_s \le 414\ \text{MPa} \\
V \ge 0.63\ \text{m}^3 \\
V = \pi r_1^2(6000 - h_1 - \sqrt{r_1^2 - b_1^2}) + 2\pi h_1 b_1^2 + \pi(b_1^2 + r_1^2)\sqrt{r_1^2 - b_1^2} + \tfrac{1}{3}\pi(\sqrt{r_1^2 - b_1^2})^3,
\end{gathered}
$$

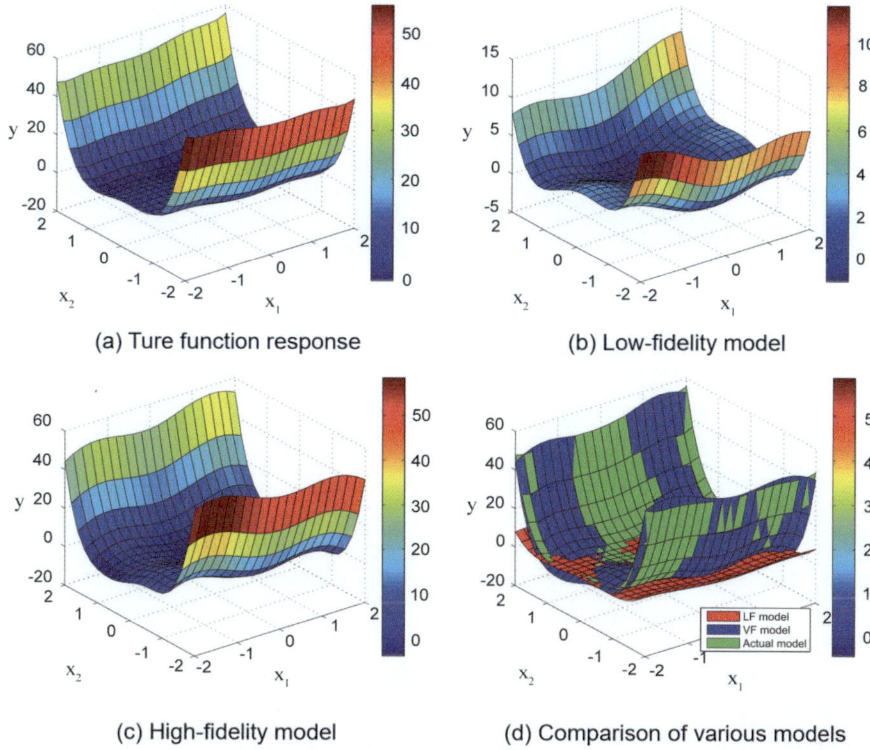

(a) Ture function response

(b) Low-fidelity model

(c) High-fidelity model

(d) Comparison of various models

Fig. 4.4 The true function response surface of function F_1 and the surrogate model $n_1 = 100$, $n_2 = 10$

$$M = \pi\left[(c_r + r_1)^2 - r_1^2\right]\left(6000 - h_1 - \sqrt{r_1^2 - b_1^2}\right) + 2\pi h_1\left[(c_b + b_1)^2 - b_1^2\right] + \frac{1}{3}\pi\left\{3\left[(c_b + b_1)^2 + (c_r + r_1)^2\right]\sqrt{(c_r + r_1)^2 - (c_b + b_1)^2} + \left[(\sqrt{(C_r + r_1)^2 - (C_b + b_1)^2}^3\right] - 3(r_1^2 + b_1^2)\sqrt{r_1^2 - b_1^2} - \left(\sqrt{r_1^2 - b_1^2}\right)^3\right\}.$$

Since the tensile stress σ_s does not have an explicit functional expression, a variable-fidelity surrogate model is needed to establish the relationship between the tensile stress and the design variables. First, the SLE-NGHS algorithm is used to generate sampling samples. Then, the tensile stress at the sampling points is simulated using Ansys 10 software. In this experiment, the high-fidelity analysis model employs an axisymmetric three-dimensional finite element simulation model with hexahedral mesh division, as shown in Fig. 4.6; the low-fidelity analysis model uses an axisymmetric one-dimensional model, as shown in Fig. 4.7.

After completing the simulation analysis of 60 low-fidelity model sampling points and 20 high-fidelity model sampling points, the variable-fidelity model establishes

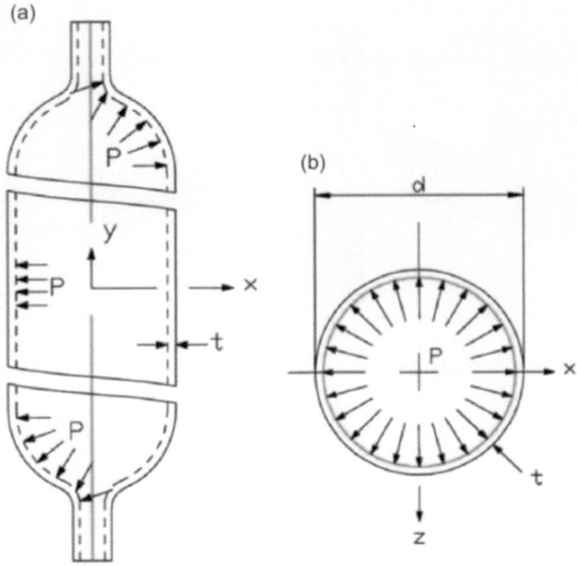

Fig. 4.5 Circular pressure vessel. **a** Pressure vessel. **b** Cross-sectional view

Table 4.4 Range of variable values

Design variable	Value range (mm)
End height h_1	200–320
End inner radius b_1	25–60
End thickness C_b	15–30
Main body inner radius r_1	150–230
Main body thickness C_r	10–23

the relationship between the tensile stress and the design variables, and 13 verification points are used to validate the prediction accuracy of the variable-fidelity surrogate model. Due to the time-consuming simulation process, the experiment is independently repeated 5 times, and the average value is taken, which is then compared with the results in the literature [11]. Table 4.5 lists the prediction accuracy indicators corresponding to different sampling methods and modeling methods. It is worth noting that the surrogate model based on the SLE-NGHS sampling method and the DM-kriging modeling method exhibits better prediction accuracy than other methods. To more intuitively demonstrate the prediction accuracy of the surrogate model based on the SLE-NGHS sampling method and the DM-kriging modeling method, Fig. 4.8 plots the corresponding relationship between the true response T (x-axis) and the predicted value P (y-axis). When the coordinates of the point (T, P) lie on the line $y = x$, it indicates that the predicted value is very close to the true response value.

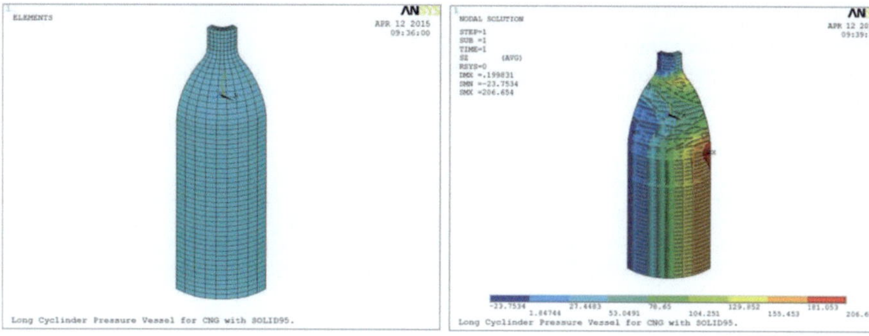

Fig. 4.6 3D high-fidelity analysis model

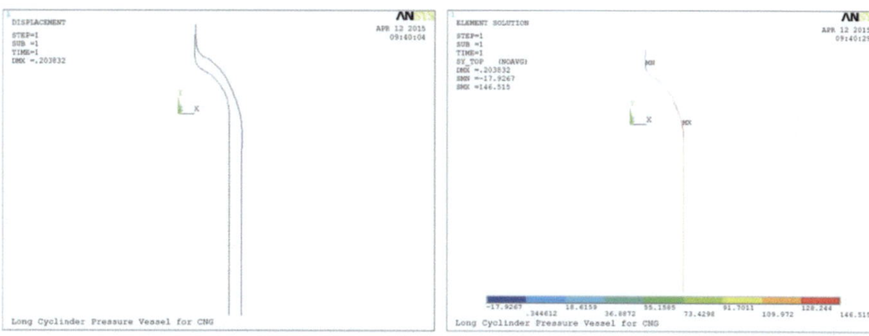

Fig. 4.7 1D low-fidelity analysis model

Table 4.5 Comparison of various accuracy indicators under different sampling methods and modeling methods

Sampling methods	Modeling methods	R^2	RRMSE	RMSE
SLE-NGHS	DM-kriging	0.9948	0.0692	3.4404
Nested	DM-kriging	0.8951	0.3129	25.6307
Nested	DM-LSSVR	0.944879	0.226818	19.22132
Nested	PKI-LSSVR	0.9625	0.187083	15.313

4.4 Summary

This chapter proposes the SLE-NGHS method, a nested LHD that uses Successive Local Enumeration for LF sampling and an improved NGHS algorithm for HF sampling. Numerical experiments and a pressure vessel case study demonstrate its superior space-filling, acceptable computational cost, and high prediction accuracy compared to Monte Carlo, nested TPLHD, and ESE-GroupRand methods. The method advances variable-fidelity surrogate modeling for complex engineering design.

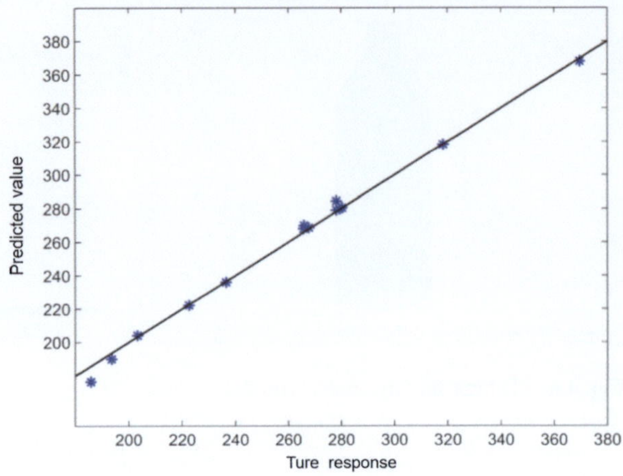

Fig. 4.8 Comparison between true circumferential stress response and predicted values of pressure vessel

References

1. Gu L (2001) A comparison of polynomial based regression models in vehicle safety analysis. In: International design engineering technical conferences and computers and information in engineering conference, vol 80227. American Society of Mechanical Engineers, pp 509–514
2. Robinson TD, Eldred MS, Willcox KE et al (2008) Surrogate-based optimization using multifidelity models with variable parameterization and corrected space mapping. AIAA J 46(11):2814–2822
3. Fang K-T, Lin DKJ, Winker P et al (2000) Uniform design: theory and application. Technometrics 42(3):237–248
4. Owen AB (1992) Orthogonal arrays for computer experiments, integration and visualization. Statistica Sinica, 439–452
5. Husslage BGM, Rennen G, van Dam ER et al (2008) Space-filling Latin hypercube designs for computer experiments (Revision of CentER DP 2006-18)
6. Zhu H, Liu L, Long T et al (2012) A novel algorithm of maximin Latin hypercube design using successive local enumeration. Eng Optim 44(5):551–564
7. Wang Y, Hao J-K, Glover F et al (2014) A Tabu search based memetic algorithm for the maximum diversity problem. Eng Appl Artif Intell 27:103–114
8. Zou D, Gao L, Wu J et al (2010) A novel global harmony search algorithm for reliability problems. Comput Indus Eng 58(2):307–316
9. Viana FAC, Venter G, Balabanov V (2010) An algorithm for fast optimal Latin hypercube design of experiments. In: Int J Numer Methods Eng 82(2):135–156
10. Rennen G, Husslage B, Van Dam ER et al (2010) Nested maximin Latin hypercube designs. Struct Multidisc Optim 41:371–395
11. Zheng J, Shao X, Gao L et al (2014) A prior-knowledge input LSSVR metamodeling method with tuning based on cellular particle swarm optimization for engineering design. Expert Syst Appl 41(5):2111–2125

Chapter 5
Variable-Fidelity Surrogate Modeling via Scale Functions

Variable-fidelity surrogate modeling integrates low-fidelity (LF) and high-fidelity (HF) models to balance computational efficiency and accuracy in multidisciplinary design optimization. After generating sample points through experimental design (e.g., the nested LHD methods in Chap. 4), HF and LF models are evaluated to obtain input-output data. Constructing an accurate surrogate model from limited HF and LF samples, while leveraging the strengths of both models, is critical. As discussed in Chap. 1, variable-fidelity approaches include scaling function-based methods, space mapping, and Co-Kriging (e.g., Hierarchical Kriging). Recent advancements, such as transfer learning, have relaxed the traditional requirement for nested designs ($X_H \subset X_L$), enabling more flexible sampling strategies [1].

This chapter proposes a scaling function-based method, the Difference Mapping Framework with Least Squares Support Vector Regression (DMF-LSSVR), which constructs a surrogate model by scaling an LF basis function and mapping the discrepancy to the HF model using LSSVR. The method's performance is validated through three engineering case studies: a deep-hole model, a vortex generator on a reverse wing, and a heat exchanger.

5.1 Scale Function Based on LSSVR

This chapter presents a variable-fidelity surrogate modeling method based on an LSSVR difference mapping framework, specifically the LSSVR with Difference Mapping Framework (DMF-LSSVR). The detailed construction workflow of DMF-LSSVR will be elaborated below, covering three core phases: Difference Mapping, Basis Function Establishment, and LSSVR Mapping.

J. Yi et al., *Variable-Fidelity Surrogate*, Engineering Applications of Computational Methods 25, https://doi.org/10.1007/978-981-95-5527-7_5

5.1.1 Gap Mapping

Figure 5.1 illustrates the difference gap-mapping workflow proposed in this study. Unlike traditional single-fidelity surrogate models, the sample points are derived from experimental design (DOE) on both high-fidelity (HF) and low-fidelity (LF) analytical models. Using these sample datasets, a surrogate basis model is first created. This basis model is then directly mapped to the high-fidelity model. If the mapped basis model passes model validation, it serves as a qualified surrogate model to approximate the high-fidelity analytical model.

5.1.2 Experiment Design for Gap Mapping

Unlike a single-fidelity surrogate, a gap-mapping-oriented experimental design is a two-level setup corresponding to low-fidelity and high-fidelity analysis models. The number of high-fidelity sample points is smaller than that of low-fidelity sample points because low-fidelity simulations are relatively less computationally expensive and easier to obtain. Given the same time and cost, the low-fidelity analysis model can be run more times.

The low-fidelity dataset is denoted by $D_L = \{(x_{1l}, y_{1l}), (x_{2l}, y_{2l}), \ldots, (x_{nl}, y_{nl})\}$ where $X_L = \{x_{1l}, x_{2l}, \ldots, x_{nl}\}$ represents the n low-fidelity input sample points, and $Y_L = \{y_{1l}, y_{2l}, \ldots, y_{nl}\}$ represents the responses of the low-fidelity analysis model at those points. Similarly, $D_H = \{(x_{1h}, y_{1h}), (x_{2h}, y_{2h}), \ldots, (x_{mh}, y_{mh})\}$, with $X_H = \{x_{1h}, x_{2h}, \ldots, x_{mh}\}$, $Y_H = \{y_{1h}, y_{2h}, \ldots, y_{mh}\}$ denotes the high-fidelity sample dataset.

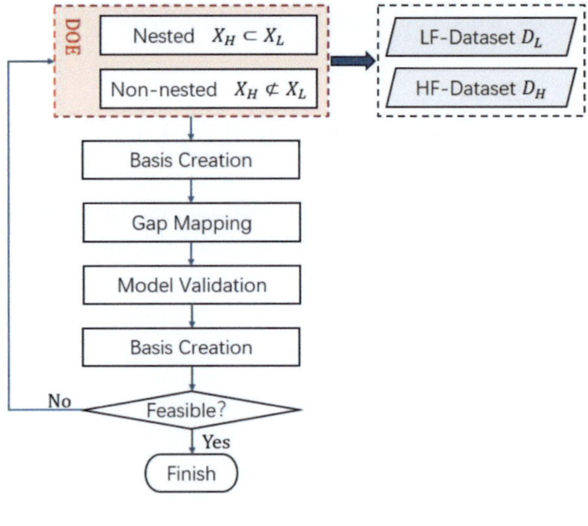

Fig. 5.1 Gap mapping workflow

This study also investigates the effect of the relationship between high- and low-fidelity samples (i.e., nested versus non-nested) on prediction accuracy. When the high-fidelity sample points are covered by the low-fidelity sample points, i.e., $X_H \subset X_L$, the samples are nested; otherwise, they are non-nested.

$$\begin{cases} X_H \subset X_L & \text{nested DOE} \\ X_H \not\subset X_L & \text{non-nested.} \end{cases} \tag{5.1}$$

5.1.3 Creation of Basis Functions

This section describes the process of constructing the basis functions shown in the flowchart Fig. 5.1. When the low-fidelity analysis model is an explicit function—such as an empirical engineering formula—the low-fidelity model $y_l(x)$ can be used directly as a basis function. Basis functions can also be obtained by approximating the low-fidelity analysis model using methods like Kriging or response-surface models. In this paper, to fully exploit the low-fidelity surrogate model and filter out some noise, the basis functions are generated by stretching the low-fidelity surrogate and can be defined by the following formula:

$$f_{base}(x) = \text{scaled}(\hat{y}_l(x)) = \rho_0 + \rho_1 \hat{y}_l(x). \tag{5.2}$$

Define two scaling factors for the basis functions, ρ_0 and ρ_1, which obtained by using the least squares method.

$$\begin{aligned} \min \quad L(\rho_0, \rho_1) = \sum_{i=1}^{m} \left[\left(\rho_0 + \rho_1 \hat{y}_l(x_i) \right) - y_h(x_i) \right]^2, \\ \text{s.t.} \quad l_0 \le \rho_0 \le u_0, \quad l_1 \le \rho_1 \le u_1, \\ x_i \in X_H. \end{aligned} \tag{5.3}$$

5.1.4 LSSVR Mapping

In the process of constructing the basis functions described above, the low-fidelity analysis model is stretched toward the high-fidelity analysis model, but there remains a tolerance (i.e., discrepancy) between the basis functions and the high-fidelity analysis model. In this subsection, LSSVR is used to approximate the discrepancy between the two, thereby creating the gap-mapping model.

First, using the stretching factors $\hat{\rho}_0$ and $\hat{\rho}_1$ from the previous subsection, compute the tolerance (discrepancy) between the basis function and the high-fidelity analysis model as $\delta = \{\delta(x_1), \delta(x_2), \ldots, \delta(x_m)\}$.

$$\delta(x_i) = y_{ih} - f_{base}(x_i) = y_{ih} - (\hat{\rho}_0 + \hat{\rho}_1 \hat{y}_l(x_i))x_i \in X_H. \tag{5.4}$$

According to a specific experimental design methodology, generate a set of sample data x_i and the corresponding tolerance responses $\delta(x_i)$. The gap-mapping model created using LSSVR can be expressed by the following equation:

$$\min \quad \frac{1}{2}\|w\|^2 + \frac{C}{2}\sum_{i=1}^{m} e_i^2 \text{s.t.} \quad y_{ih} - \delta(x_i) - w^T \cdot \varphi(x_i) - b = e_i i = 1, 2, \ldots, m, \tag{5.5}$$

where m is the number of HF sample points.

By introducing the Lagrangian transformation, the above equation can be expressed as

$$L = \frac{1}{2}\|w\|^2 + \frac{1}{2}C\sum_{i=1}^{m} e_i^2 - \sum_{i=1}^{m} \alpha_i \left(y_i - w^T\varphi(x_i) - b - e_i\right). \tag{5.6}$$

According to the KKT conditions, taking the partial derivatives concerning the Lagrangian multipliers in the above equation yields the following equation:

$$\begin{aligned} \partial_b L = \sum_{i=1}^{m} \alpha_i = 0, \partial_w L = w - \sum_{i=1}^{m} \alpha_i \varphi(x_i) = 0, \partial_{e_i} L = \alpha_i - Ce_i = 0, \\ \partial_{\alpha_i} L = w^T \varphi(x_i) + b + e_i - y_i = 0. \end{aligned} \tag{5.7}$$

After eliminating the variables $\mathbf{w}$ and e_i, the optimization problem in the above equation can be expressed as

$$\begin{bmatrix} 0 & e^\tau \\ e & \mathbf{K} + C^{-1}\mathbf{I} \end{bmatrix} \begin{bmatrix} b \\ a \end{bmatrix} = \begin{bmatrix} 0 \\ \delta, \end{bmatrix} \tag{5.8}$$

where $\mathbf{a} = [\alpha_1, \alpha_2, \ldots, \alpha_m]^T$; $\mathbf{e}$ is an all-ones vector expressed as $I^T = [1, \ldots, 1]_{1\times m}$; $\mathbf{K}$ is an $m \times m$ matrix whose elements are given by $K_{ij} = \varphi\left(\left[X_{hi} Y_{lj}^h\right]\right)^T \varphi\left(\left[X_{lj} Y_{lj}^h\right]\right) = k\left(\left[X_{hi} Y_{hi}^h\right], \left[X_{hj} Y_{lj}^h\right]\right)$, where $k\left(\left[X_{hi} Y_{li}^h\right],\right.$ $\left.\left[X_{hj} Y_{lj}^h\right]\right)$ is the kernel function.

For kernel functions, the commonly used ones mainly include linear kernel functions, polynomial kernel functions, sigmoid kernel functions, Gaussian kernel functions, and so on, which are, respectively, expressed as follows:

linear kernel functions:

$$k(\mathbf{x}, \mathbf{x}') = \mathbf{x}^T \mathbf{x}'$$

polynomial kernel functions:

$$k(\mathbf{x}, \mathbf{x}') = (\mathbf{x} \cdot \mathbf{x}')^d$$

Gaussian kernel functions:

$$k(\mathbf{x}, \mathbf{x}') = \exp\left(-\frac{\|\mathbf{x} - \mathbf{x}'\|^2}{2\sigma^2}\right)$$

sigmoid kernel functions:

$$k(\mathbf{x}, \mathbf{x}') = \tanh\left(\kappa \langle \mathbf{x} \cdot \mathbf{x}' \rangle + \nu\right)$$

Compared to other kernel functions, the Gaussian kernel has only the parameter σ, fewer parameters overall, and better performance, making it the preferred choice in most cases. Therefore, in the DMF-LSSVR construction process, the Gaussian kernel is selected and expressed as follows:

$$k\left(\left[\mathbf{X}_{hi}, \mathbf{Y}_{li}^h\right], \left[\mathbf{X}_{lj}, \mathbf{Y}_{lj}^h\right]\right) = \exp\left(-\frac{\left\|\left[\mathbf{X}_{hi}, \mathbf{Y}_{li}^h\right] - \left[\mathbf{X}_{lj}, \mathbf{Y}_{lj}^h\right]\right\|^2}{2\sigma^2}\right). \tag{5.9}$$

After obtaining the solutions for α and b, the DMF-LSSVR model can be expressed as

$$\hat{f}(x) = f_{\text{base}}(x) + \hat{\delta}(x) = \hat{\rho}_0 + \hat{\rho}_1 \hat{y}_l(x) + \sum_{i=1}^{m} \alpha_i K((\mathbf{x} \cdot \mathbf{x}_i)) + b. \tag{5.10}$$

5.2 Examples and Analysis

In this subsection, three different engineering case studies (namely, the deep-hole model, the design of a vortex generator on a reverse wing, and the design of a heat exchanger) are used to compare the prediction accuracy of the proposed DMF-LSSVR model with that of a conventional single-fidelity surrogate model and a typical linearly corrected multi-fidelity surrogate model [2], thereby validating the performance of the proposed method. Three different accuracy metrics are adopted, and the influence of nested versus non-nested experimental design relationships on prediction accuracy is also examined. The three accuracy metrics-root mean square error (RMSE), maximum absolute error (MAX), and the coefficient of determination (R-squared)—are used to evaluate the performance of the proposed multi-fidelity surrogate model.

5.2.1 Deep Hole Model

In this subsection, the borehole model (Chap. 3 see Sect. 3.2) is used to validate the proposed multi-fidelity surrogate model, while also considering the effect of different experimental design sizes on prediction accuracy. At the same time, for a given sample size, both nested and non-nested experimental designs are employed to explore how the relationship between high- and low-fidelity samples affects surrogate accuracy. The comparative prediction accuracy values for the different surrogate techniques are shown in Table 5.1, where DMF-LSSVR denotes the gap-mapping model based on least squares support vector regression proposed in this paper, and VFP denotes the linearly corrected model from [2]. Both of these multi-fidelity methods use high- and low-fidelity sample data, whereas the single-fidelity model uses only high-fidelity samples. In the table, "Nested" refers to nested samples and "Non-nested" refers to non-nested samples.

From the Table 5.1, it can be seen that as the number of sample points increases, the prediction accuracy of all models improves. However, the DMF-LSSVR model performs the best. The DMF-LSSVR model with only 10 high-fidelity (HF) sample points achieves higher accuracy than the other models with 50 sample points. Additionally, the table shows that model accuracy under a nested design is higher than under a non-nested design, indicating that nested experimental design helps the multi-fidelity surrogate model improve its prediction accuracy.

To more intuitively illustrate the surrogate performance of each model, Fig. 5.2 shows the predicted response values from the different surrogate models. The horizontal axis represents the true values, and the vertical axis represents the predicted values. The closer a red "*" is to the line $y = x$, the smaller its prediction error; when a red "*" lies exactly on this line, the prediction error is zero. It can be seen that the DMF-LSSVR model under the nested experimental design has the smallest prediction error. Figures 5.3, 5.4 and 5.5 display line plots of RMSE, MAX, and R-square

Table 5.1 Comparison results of prediction accuracy for the deep-hole model under different sample sizes

Sample size	DMF-LSSVR		VFP	Kriging	RSM	RBF	
	Nested	Non-nested					
10	RMSE	2.9013	5.4683	53.10624	36.21304	114.405	64.32574
	MAX	9.225	15.7264	97.20619	103.6399	212.0382	152.5603
	R-square	0.9971	0.9898	0.038786	0.552554	−3.3883	−0.35917
20	RMSE	2.7885	4.6383	86.21786	26.13535	189.4494	59.97801
	MAX	8.7586	15.2258	260.999	89.01907	575.9454	105.4478
	R-square	0.9973	0.9926	−1.44242	0.766802	−8.82643	−0.21906
50	RMSE	1.3915	3.2621	79.65192	13.47673	14.83255	60.97156
	MAX	5.1683	12.1424	19.15665	28.07668	33.18127	133.2916
	R-square	0.9993	0.9964	0.978257	0.937784	0.92461	−0.27388

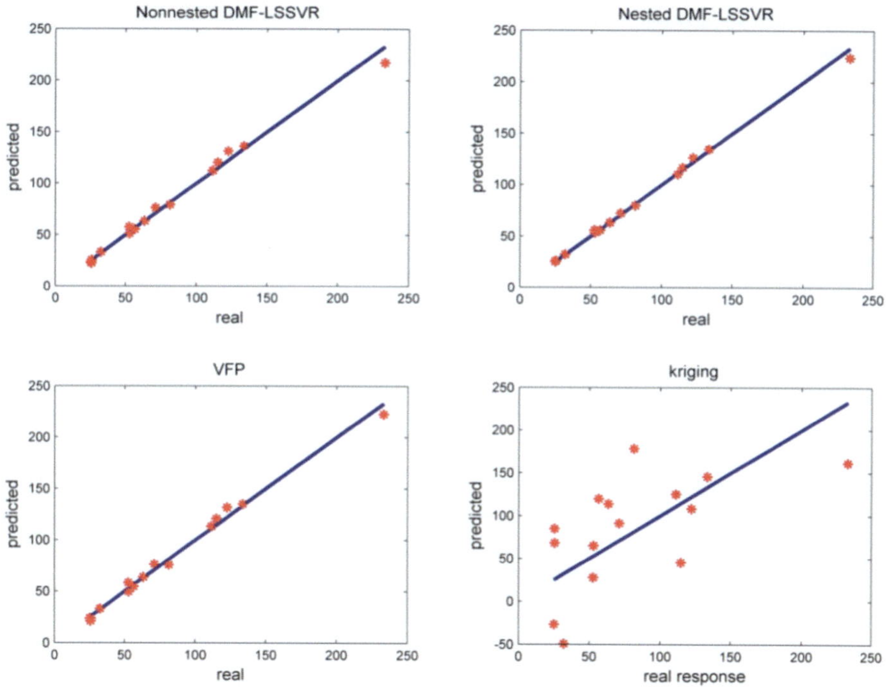

Fig. 5.2 Predicted response values of different surrogate models

Fig. 5.3 RMSE for different surrogate models

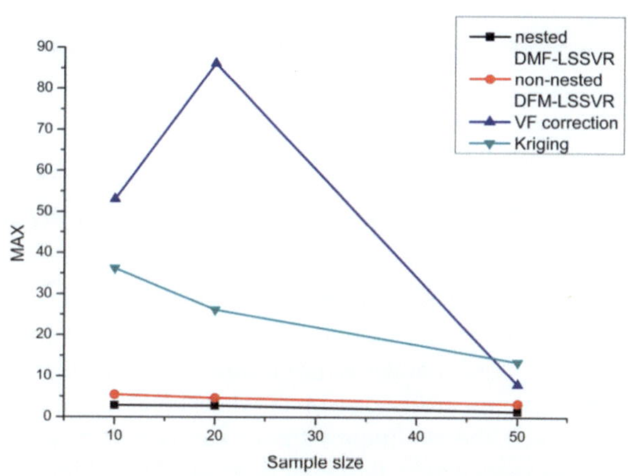

for the various surrogate models under different sample sizes. In the first two plots, the DMF-LSSVR line under nested design lies lowest, and in the R-square plot, DMF-LSSVR lies highest. This demonstrates that, under the nested experimental design, DMF-LSSVR achieves the best predictive accuracy.

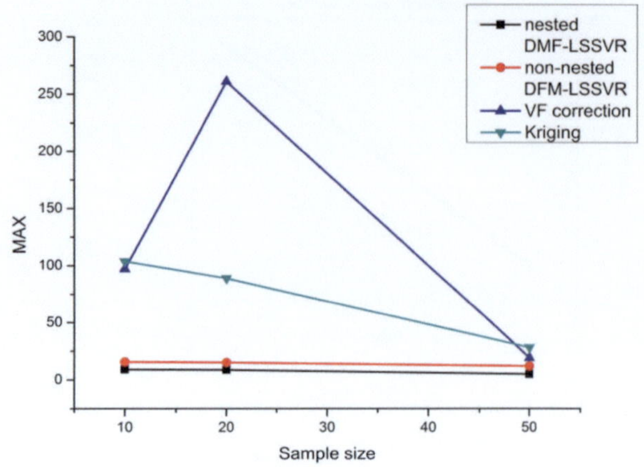

Fig. 5.4 MAX for different surrogate models

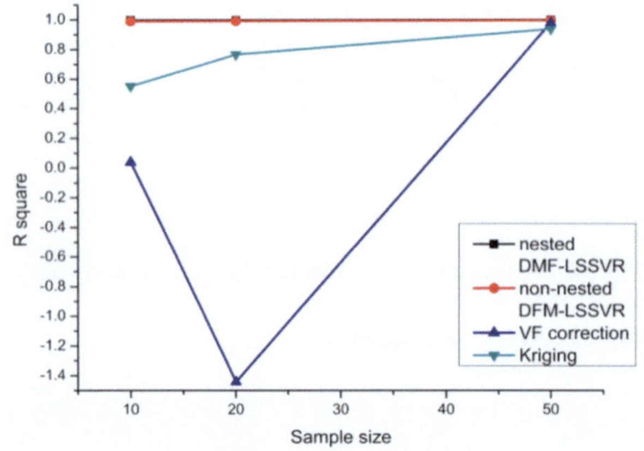

Fig. 5.5 R-squared values for different surrogate models

5.2.2 *Design of Vortex Generators for Reverse Wings*

When a moving aircraft flies very close to the ground (or water surface), the pressure difference between its upper and lower surfaces increases, causing the lift to rise sharply—this is known as the ground effect. Thanks to this ground-effect mechanism, inverted wings are used on open-wheel race cars, such as F1 and Indy cars, to generate downforce, thereby improving the car's traction and cornering speed [3–5].

In this subsection, the design data of a 2 mm-high inverted rotating-joint boundary-layer vortex generator device, as described by Kuya [6], are used to evaluate the performance of the proposed method. In this context, the wind-tunnel test data serve as high-fidelity (HF) data, while the simulation results are treated as low-fidelity (LF) information. The HF experimental data are shown in Table 5.2, with design variables including wing angle of attack, chassis height, and downforce response. The HF response is obtained by running wind-tunnel tests at the design points, whereas the

Table 5.2 High-reliability experimental data

Coded design variables		HF response
$X_1(=\alpha)$	$X_2(=h/c)$	
0	0	2.46
0.5	0	1.79
1	0	1.57
0	0.5	1.81
0.5	0.5	1.96
1	0.5	1.33
0	1	1.46
0.5	1	1.8
1	1	1.31

Table 5.3 9 low-fidelity (LF) points for nested and non-nested designs

Nested design			Non-nested design		
$X_1(=\alpha)$	$X_2(=h/c)$	Y_1	$X_1(=\alpha)$	$X_2(=h/c)$	Y_1
0	0	2.29	0.72	0.06	1.69
0.5	0	1.75	0.39	0.17	1.86
1	0	1.53	0.06	0.28	2.17
0	0.5	1.75	0.61	0.39	1.94
0.5	0.5	1.98	0.94	0.50	1.80
1	0.5	1.78	0.17	0.61	2.22
0	1	1.43	0.5	0.72	2.28
0.5	1	2.57	0.83	0.83	1.94
1	1	1.81	0.28	0.94	2.28

LF response is obtained via three-dimensional steady Reynolds-averaged Navier-Stokes (RANS) simulations [6]. Both nested and non-nested experimental design relationships are considered, and the effect of different LF sample sizes on surrogate performance is also investigated. For different LF sample sizes, the nested and non-nested experimental designs are given in Tables 5.3 and Table 5.4, respectively.

In Table 5.5, the observed and true values at four validation points are presented, along with the corresponding prediction-accuracy metrics. From the table, it can be seen that the LF sample size also affects the accuracy of the final predictive model. As the number of LF sample points increases, the prediction accuracy of the resulting multi-fidelity model also improves. Furthermore, the proposed DMF-LSSVR model outperforms Kuya's Cokriging model in terms of prediction accuracy. For both DMF-LSSVR and Cokriging, nested experimental designs generally yield higher prediction accuracy than non-nested designs.

Table 5.4 25 low-fidelity (LF) points for nested and non-nested designs

Nested design			Non-nested design		
$X_1(=\alpha)$	$X_2(=h/c)$	Y_L	$X_1(=\alpha)$	$X_2(=h/c)$	Y_L
0	0	2.29	0.5	0.02	1.77
0.25	0	1.83	0.14	0.06	2.52
0.5	0	1.75	0.42	0.1	1.84
0.75	0	1.62	0.06	0.14	2.33
1	0	1.53	0.18	0.18	2.56
0	0.25	2	0.62	0.22	1.87
0.25	0.25	2.59	0.74	0.26	1.83
0.5	0.25	1.91	0.86	0.3	1.78
0.75	0.25	1.82	0.38	0.34	2.22
1	0.25	1.7	0.58	0.38	1.95
0	0.5	1.75	0.22	0.42	2.51
0.25	0.5	2.52	0.34	0.46	2.63
0.75	0.5	1.92	0.54	0.54	1.98
1	0.5	1.78	0.3	0.58	2.57
0	0.75	1.57	0.9	0.62	1.85
0.25	0.75	2.34	0.02	0.66	1.69
0.5	0.75	2.36	0.82	0.7	1.92
0.75	0.75	1.98	0.26	0.74	2.38
1	0.75	1.81	0.7	0.78	2.01
0	1	1.43	0.46	0.82	2.59
0.25	1	2.19	0.98	0.86	1.82
0.5	1	2.57	0.1	0.9	1.8
0.75	1	2.01	0.78	0.94	1.97
1	1	1.81	0.66	0.98	2.05

The observed values at the validation points for the different surrogate models are shown in Fig. 5.6. It can be seen that under the nested design, the DMF-LSSVR model is closest to the diagonal line, indicating that the DMF-LSSVR model has the highest accuracy when using a nested experimental design.

5.2.3 *Heat Exchanger*

In this subsection, a typical electronic cooling-system heat exchanger design problem is considered. As shown in Fig. 5.7, this is a heat-dissipation device for heat sources such as microprocessors. The factors affecting heat-dissipation performance are the inlet temperature T_{in}, the total mass-flow rate m, the heat-source temperature T_{wall},

Table 5.5 Model validation data

Approach	LF sample size	Sampling strategy	$(X_1(=a),\ X_2(=h/c))$				RMSE	MAX	R-square
			(0.25, 0.25)	(0.75, 0.25)	(0.25, 0.75)	(0.75, 0.75)			
True (experimental)	–	–	2.42	1.87	2.60	1.75	–	–	–
DMF-LSSVR	9	Nested	1.9499	1.5879	1.9205	1.8586	0.4899	0.6195	0.1861
		Non-nested	1.9751	1.6107	1.8644	1.6707	0.4507	0.7356	0.2548
	25	Nested	2.6426	1.7199	2.2584	1.6683	0.2211	0.3416	0.6408
		Non-nested	2.4415	1.8591	2.4318	1.9669	0.1378	0.2169	0.8524
Cokriging	9	Nested	1.99	1.64	1.74	1.61	0.4992	0.8600	0.1708
		Non-nested	2.03	1.60	1.79	1.55	0.4799	0.8100	0.2392
	25	Nested	2.37	1.69	1.98	1.55	0.3389	0.6200	0.4179
		Non-nested	2.51	1.56	2.08	1.51	0.3287	0.5200	0.4264
Cokriging Interpretation	25	Nested	2.48	1.69	1.98	1.48	0.3512	0.6200	0.3580
		Non-nested	2.65	1.57	2.09	1.48	0.3449	0.5100	0.3143

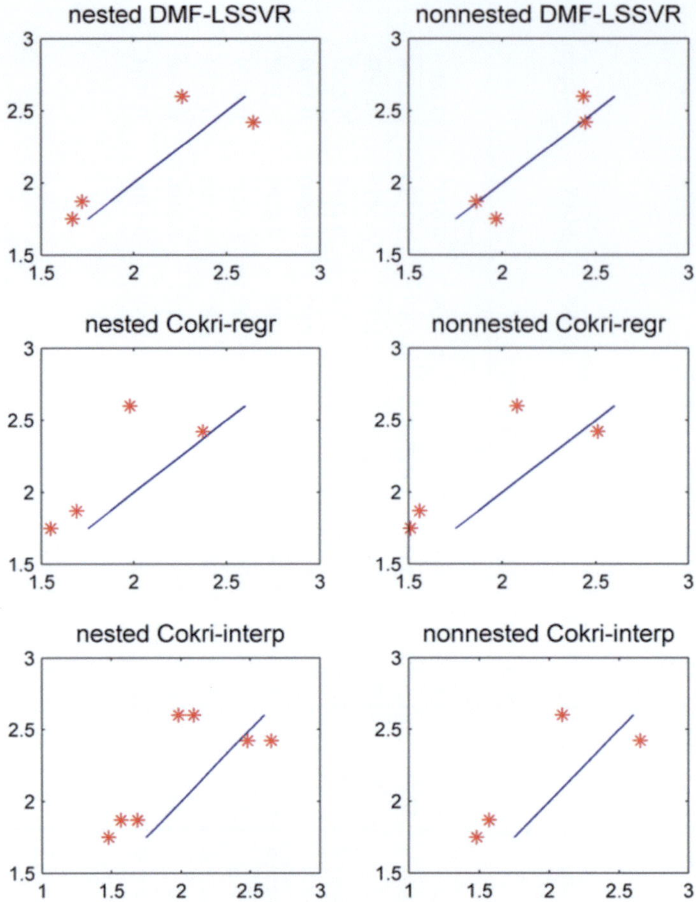

Fig. 5.6 Prediction errors of validation points for different surrogate models

and the thermal conductivity k of the solid material. The design objective is to achieve the maximum steady-state heat-conduction rate. More details can be found in [7].

Two types of simulation-FLUENT simulation (HF model) and the relatively low-cost finite-difference simulation (LF)-are both used to analyze the effects of the above factors on heat-conduction rate. The finite-difference method is a numerical simulation technique for solving two or three-dimensional heat-conduction problems [8]. The FLUENT simulation is based on partial differential equations and the finite-volume method, making it more accurate and reliable than the finite-difference method. Compared to the finite-difference method, FLUENT requires 2–3 times more simulation computation time. However, the simulation accuracy of FLUENT is also more than 10–15% higher. Table 5.6 lists the LF samples generated by a Latin hypercube based on an orthogonal array [7], where represents the finite-difference

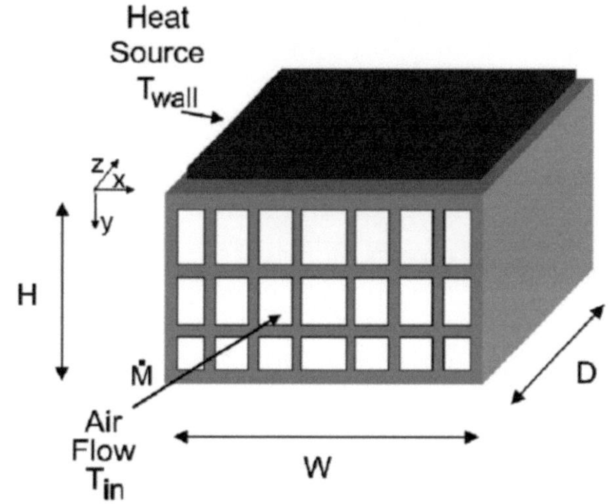

Fig. 5.7 Compact forced convection heat exchanger

Table 5.6 Sample points of finite-difference simulation [7]

Run	m	T_{in}	k	T_{wall}	Y_t
1	0.000552	293.53	318.63	388.29	25.61
2	0.000557	290.18	298.27	377.49	23.24
3	0.000566	285.77	266.71	367.27	21.23
4	0.000578	302.17	358.13	343.72	11.44
5	0.000580	272.26	211.71	333.65	15.03
6	0.000589	278.16	225.78	351.83	18.55
7	0.000594	279.54	258.51	360.13	20.74
8	0.000603	296.75	323.15	399.45	28.40
9	0.000612	280.83	291.53	394.72	30.22
10	0.000615	300.28	270.74	335.79	9.53
11	0.000626	284.89	305.46	352.29	18.13
12	0.000627	287.60	243.96	382.54	25.02
13	0.000639	270.45	241.21	341.81	17.92
14	0.000643	276.17	216.99	371.60	24.02
15	0.000652	298.04	303.96	361.58	17.47
16	0.000657	294.24	303.63	373.53	22.48
17	0.000669	296.33	343.16	385.81	25.07
18	0.000687	303.07	321.41	370.48	18.93
19	0.000683	287.05	227.31	358.24	18.91
20	0.000689	272.70	260.91	355.37	21.47
21	0.000694	278.35	212.79	376.24	25.11

(continued)

Table 5.6 (continued)

Run	m	T_{in}	k	T_{wall}	Y_t
22	0.000698	277.52	299.39	338.40	16.02
23	0.000711	292.26	273.31	392.54	27.47
24	0.000714	283.08	306.69	344.34	16.43
25	0.000722	276.53	353.75	374.41	26.50
26	0.000730	285.51	217.74	383.92	25.88
27	0.000738	295.01	295.02	347.22	14.37
28	0.000741	270.95	275.19	356.87	22.36
29	0.000751	287.99	232.02	354.08	18.17
30	0.000757	300.64	253.05	391.68	14.37
31	0.000763	292.82	254.84	373.38	21.96
32	0.000772	278.93	301.75	331.55	14.02
33	0.000782	299.86	317.84	348.41	13.68
34	0.000786	275.51	247.29	340.19	16.82
35	0.000791	271.64	284.88	365.09	25.06
36	0.000800	291.42	341.48	358.59	18.83
37	0.000803	281.47	232.64	389.46	28.69
38	0.000814	286.39	339.92	332.40	12.68
39	0.000823	288.53	207.55	393.49	27.96
40	0.000828	297.33	280.13	379.86	23.17
41	0.000836	289.62	347.65	335.44	12.79
42	0.000842	294.39	203.45	346.05	17.75
43	0.000851	273.71	315.27	381.14	29.08
44	0.000857	282.12	262.30	350.10	18.25
45	0.000865	274.35	335.16	362.30	23.89
46	0.000870	295.76	237.65	366.25	19.36
47	0.000874	282.50	253.25	396.36	30.90
48	0.000882	299.22	288.45	385.07	24.45
49	0.000891	273.43	336.04	386.95	31.05
50	0.000901	302.02	249.57	382.33	22.64
51	0.000903	284.25	290.90	364.99	22.22
52	0.000911	280.17	355.34	370.03	25.03
53	0.000922	276.89	310.73	397.78	33.27
54	0.000929	298.65	205.40	349.02	13.67
55	0.000934	288.86	265.53	339.54	13.89
56	0.000943	292.77	231.01	330.19	10.16
57	0.000947	283.62	222.95	378.66	25.48
58	0.000956	290.33	312.97	368.96	22.62
59	0.000964	271.23	348.00	398.52	35.05
60	0.000968	297.80	244.50	337.41	10.99

(continued)

Table 5.6 (continued)

Run	m	T_{in}	k	T_{wall}	Y_t
61	0.000979	291.21	283.10	353.60	17.45
62	0.000985	301.50	220.37	363.20	17.14
63	0.000987	281.11	329.45	342.32	16.95
64	0.000996	275.01	278.27	390.35	31.35

Table 5.7 22 points of FLUENT simulation [7]

Run	m	T_{in}	k	T_{wall}	Y_t	Y_s
1	0.000552	293.53	318.63	388.29	25.61	23.54
2	0.000566	285.77	266.71	367.27	21.23	20.15
3	0.000578	302.17	358.13	343.72	11.44	10.17
4	0.00058	272.26	211.71	333.65	15.03	15.29
5	0.000589	278.16	225.78	351.83	18.55	18.39
6	0.000594	279.54	258.51	360.13	20.74	20.52
7	0.000612	280.83	291.53	394.72	30.22	30.12
8	0.000626	284.89	305.46	352.29	18.13	18.13
9	0.000627	287.6	243.96	382.54	25.02	25.08
10	0.000639	270.45	241.21	341.81	17.92	19.05
11	0.000643	276.17	216.99	371.6	24.02	24.96
12	0.000652	298.04	303.96	361.58	17.47	16.95
13	0.000657	294.24	303.63	373.53	22.48	22.3
14	0.000751	287.99	232.02	354.08	18.17	19.57
15	0.000763	292.82	254.84	373.38	21.96	23.33
16	0.000814	286.39	339.92	332.4	12.68	14.36
17	0.000842	294.39	203.45	346.05	17.75	15.12
18	0.000851	273.71	315.27	381.14	29.08	34.8
19	0.000857	282.12	262.3	350.1	18.25	21.31
20	0.000874	282.5	253.25	396.36	30.9	36.11
21	0.000882	299.22	288.45	385.07	24.45	27.36
22	0.000903	284.25	290.9	364.99	22.22	25.37

response. Among these 64 low-fidelity samples, 22 sample points were selected to run FLUENT simulations, as shown in Table 5.7.

Using the above data, the DMF-LSSVR model was constructed. An additional 14 test points were used to validate this model. At these 14 points, both FLUENT and finite-difference simulation models were run to obtain the corresponding high- and low-fidelity responses. At the same time, the proposed model is compared with Kriging, response-surface method (RSM), radial basis function (RBF), and Qian-s method [7], as shown in Table 5.8. "$\hat{Y}_{\text{DMF-LSSVR}}$" denotes the proposed method, while "$\hat{Y}_{\text{ini}}$," "$\hat{Y}_{\text{RBF}}$," "$\hat{Y}_{\text{RSM}}$," and "Qian" denote Kriging, RBF, RSM, and Qian's method [7], respectively.

Table 5.8 Test comparison results of different methods

m	T_{in}	k	T_{wall}	Y_h	$\hat{Y}_{DN\text{-}FSSVR}$	$\hat{Y}_{Kri}$	$\hat{Y}_{RBF}$	$\hat{Y}_{RSM}$	Qian
0.0005	293.15	362.73	393.15	25.82	27.51	44.8	27.7278	23.8	23.85
0.00055	315	310	365	7.48	12.64	22.6	10.9606	12.1154	10.31
0.00056	277.01	354.98	374	19.77	20.00	42.16	28.9546	16.6298	16.62
0.00062	275	225	340	18.78	16.06	40.83	15.9257	16.4524	16.64
0.00068	313.28	259.12	350	4.55	10.37	23.26	21.8427	10.7537	6.44
0.00007	288.15	300	400	34.45	30.86	40.83	28.4575	30.8328	31.93
0.00008	292.73	267.84	369	21.97	21.53	44.4	21.2237	21.8264	23.7
0.00008	303.15	250	350	14.83	13.62	38.26	21.7274	14.141	6.34
0.00085	270	325	385	32.85	36.17	49.01	36.8737	37.1172	37.88
0.00085	301.31	317.85	341	11.92	12.69	39.34	8.5041	12.4922	12.99
0.00091	248.87	206.74	398	47.05	44.12	22.15	50.1849	52.876	51.77
0.00094	271.32	362.73	400	42.93	42.16	47.80	35.8777	43.204	44.97
0.00095	280	270	330	17.41	17.07	34.93	22.4704	16.4173	16.82
0.0001	293.15	202.4	273.15	22.89	18.84	21.41	11.5515	−6.5977	25.74

Table 5.9 Comparison of prediction accuracy for different models

Error criteria	DMF-LSSVR	Kriging	RBF	RSM	Qian
RMSE	3.4176	18.23015	8.382678	6.990122	3.795
MAX	7.2395	27.42	29.4877	17.2927	8.49
R-square	0.9144	−0.0845	0.486083	0.645996	0.9017

In addition, the RMSE, MAX, and R-squared values for the different modeling methods were calculated, and the results are shown in Table 5.9. It can be seen that, compared to the other models, the proposed DMF-LSSVR method achieves a more accurate surrogate.

5.3 Summary

This chapter introduces the DMF-LSSVR method, a scaling function-based approach for variable-fidelity surrogate modeling. It constructs a scaled LF basis function and uses LSSVR to map the discrepancy to HF responses. Validated through three engineering case studies, DMF-LSSVR outperforms single-fidelity models (Kriging, RSM, RBF) and other multi-fidelity methods, achieving superior accuracy with fewer HF samples. Nested designs generally yield higher accuracy than non-nested ones. The method's integration with experimental designs like SLE-NGHS (Chap. 4) enhances its applicability to complex engineering problems.

References

1. Tian K, Li Z, Zhang J et al (2021) Transfer learning based variable-fidelity surrogate model for shell buckling prediction. In: Composite structures, vol 273, p 114285 (2021)
2. Zadeh PM, Toropov VV, Wood AS (2009) Metamodel-based collaborative optimization framework. In: Structural and multidisciplinary optimization, vol 38, pp 103–115 (2009)
3. Katz J (2006) Aerodynamics of race cars. Ann Rev Fluid Mech 38(1):27–63 (2006)
4. Zhang X, Toet W, Zerihan J (2006) Ground effect aerodynamics of race cars. Appl Mech Rev 59:33 (2006)
5. Kuya Y, Takeda K, Zhang X (2010) Computational investigation of a race car wing with vortex generators in ground effect. J Fluids Eng 132(2) (2010)
6. Kuya Y, Takeda K, Zhang X et al (2011) Multifidelity surrogate modeling of experimental and computational aerodynamic data sets. AIAA J 49(2):289–298 (2011)
7. Qian Z, Seepersad CC, Roshan Joseph V et al (2006) Building surrogate models based on detailed and approximate simulations. J Mech Des 128(4):668–677
8. Incropera FP, De Witt DP, Bergman TL, Lavine AS (1985) Fundamentals of heat and mass transfer. Wiley, Hoboken, pp 939–940

Chapter 6
Variable-Fidelity Physics-Informed Neural Networks

Physics-Informed Neural Networks (PINNs) integrate physical laws into neural network architectures, offering a hybrid approach to solve partial differential equations (PDEs) with high accuracy and computational efficiency. However, standard PINNs often rely on abundant high-fidelity data or precise low-fidelity models, which can be limiting in scenarios with sparse data or complex physics. The Variable-Fidelity Physics-Informed Neural Network (VF-PINN) framework addresses these challenges by combining low-fidelity analytical models with sparse high-fidelity data, achieving a balance between accuracy and computational cost. This chapter proposes a generalized VF-PINN framework and demonstrates its applicability through two case studies: Euler-Bernoulli beam with variable point load, and two-dimensional Poisson equation with variable source intensity. Each case study highlights the framework's ability to handle diverse physical systems, providing a comparative analysis of performance against traditional methods.

6.1 Physics-Informed Neural Networks

PINNs provide a powerful framework for solving PDEs by embedding physical laws, expressed as governing equations, and their associated boundary conditions into the loss function of a neural network [1]. This approach ensures that the trained network not only approximates real physical data but also adheres to properties such as symmetry, invariance, and conservation inherent in the PDEs. Consider a general PDE with parameter λ for the solution $u(\mathbf{x})$, defined on a domain $\Gamma \subset \mathbb{R}$ with spatial coordinates $\mathbf{x} = (x_1, x_2, \ldots, x_n)$:

$$f\left(\mathbf{x}; \frac{\partial u}{\partial x_1}, \ldots, \frac{\partial^2 u}{\partial x_1 \partial x_n}; \lambda\right) = 0, \quad \mathbf{x} \in \Gamma, \tag{6.1}$$

J. Yi et al., *Variable-Fidelity Surrogate*, Engineering Applications of Computational Methods 25, https://doi.org/10.1007/978-981-95-5527-7_6

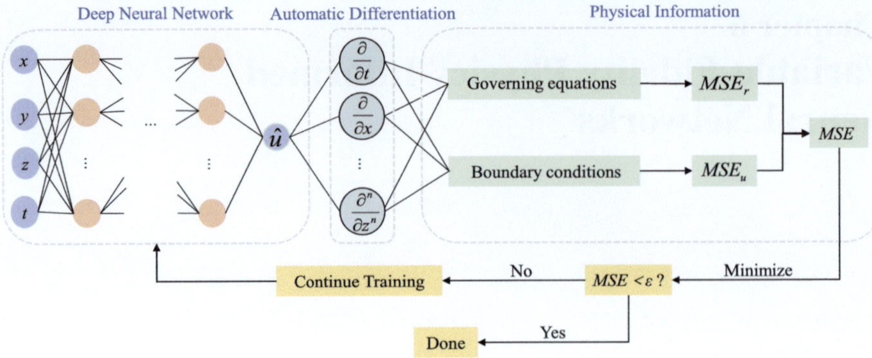

Fig. 6.1 Network structure of PINN

subject to boundary conditions $u(\mathbf{x}) = b_0(\mathbf{x})$ on $\partial\Gamma$ and initial conditions where applicable. Here, f represents the PDE expression, involving differential operators and parameters $\lambda = (\lambda_1, \lambda_2, \ldots, \lambda_n)$, while boundary conditions may include Dirichlet, Neumann, or Robin types.

Figure 6.1 depicts the neural network structure of PINN. It consists of a fully connected feedforward neural network, taking spatial coordinates $\mathbf{x}$ as inputs and outputting an approximation of $u(\mathbf{x})$. The network comprises multiple hidden layers, defined iteratively as $Z^k(\mathbf{x}) = \Phi(w^k Z^{k-1} + b^k)$ for $1 \leq k \leq K - 1$, where Φ is a nonlinear activation function (e.g., sigmoid, tanh, or ReLU), w^k and b^k are weights and biases, and the final layer $Z^K(\mathbf{x}) = w^K Z^{K-1} + b^K$ yields the solution. Activation functions introduce nonlinearity, enabling the network to capture complex physical behaviors, though care must be taken to avoid issues like gradient vanishing or exploding during training.

The loss function, critical to PINN's design, is formulated as the mean squared error (MSE):

$$\text{MSE} = \text{MSE}_r + \text{MSE}_u, \tag{6.2}$$

where $\text{MSE}_r = \frac{1}{N_r} \sum_{i=1}^{N_r} |\hat{r} - r|^2$ enforces the PDE residual across collocation points, and $\text{MSE}_u = \frac{1}{N_u} \sum_{i=1}^{N_u} |\hat{u} - u|^2$ ensures boundary condition satisfaction. Optimization algorithms, such as Adam, iteratively update the trainable parameters w^k and b^k to minimize this loss.

A key enabler of PINN is automatic differentiation (AD), which computes the derivatives in the PDE residual (e.g., $\frac{\partial u}{\partial x_i}$) directly from the network's explicit expressions. Unlike traditional numerical methods, AD avoids discretization and truncation errors, leveraging machine learning frameworks like TensorFlow or PyTorch for efficient derivative calculations via the chain rule. This enhances the accuracy and flexibility of PINN in solving PDEs, making it a robust tool for engineering and physics applications, though its performance can depend on the quality of training data and the choice of optimizer.

6.2 Variable-Fidelity PINN

6.2.1 Motivation

In many engineering applications, such as structural mechanics, fluid dynamics, and electromagnetics, solving PDEs using high-fidelity numerical methods like the Finite Element Method (FEA) or Computational Fluid Dynamics (CFD) is computationally expensive, particularly for complex geometries or multi-scale phenomena. Standard PINNs offer a promising alternative by embedding physical laws into neural network training, yet they often require extensive high-fidelity data to achieve accurate solutions. This reliance on high-fidelity data is impractical in scenarios where such data are scarce or costly to obtain, limiting the applicability of standard PINNs in large-scale or real-time engineering problems.

The Variable-Fidelity PINN (VF-PINN) framework addresses these challenges by synergizing low-fidelity analytical solutions or simplified numerical models with sparse high-fidelity data, guided by physical constraints. This multi-fidelity approach significantly reduces computational costs while maintaining high accuracy, making VF-PINN particularly suitable for design optimization and real-time prediction tasks where rapid and reliable solutions are critical. By leveraging the efficiency of low-fidelity models and correcting their inaccuracies with limited high-fidelity information, VF-PINN achieves an effective balance between computational efficiency and solution precision.

6.2.2 Framework Architecture

The VF-PINN framework employs a multi-network architecture designed to process low- and high-fidelity data separately before integrating them to produce the final solution. It consists of three key components:

- **Low-Fidelity Network (NN1)**: A neural network that processes low-fidelity input data $\mathbf{x}_{LF}$ (e.g., outputs from analytical models or coarse-grid simulations) through multiple hidden layers with hyperbolic tangent (tanh) activation functions. The output, Y_L, represents an intermediate approximation of the low-fidelity solution.
- **High-Fidelity Network (NN2)**: A parallel neural network that handles sparse high-fidelity data $\mathbf{x}_{HF}$ (e.g., from high-resolution FEA simulations). It also employs tanh activation functions and produces an intermediate output Y_H, capturing high-fidelity corrections.
- **Fusion Network (NN_f)**: A final neural network that combines Y_L and Y_H to generate the predicted solution Y. The relationship between the high-fidelity and low-fidelity solutions is modeled as

$$Y_H = Y_L + \delta(\mathbf{x}), \tag{6.3}$$

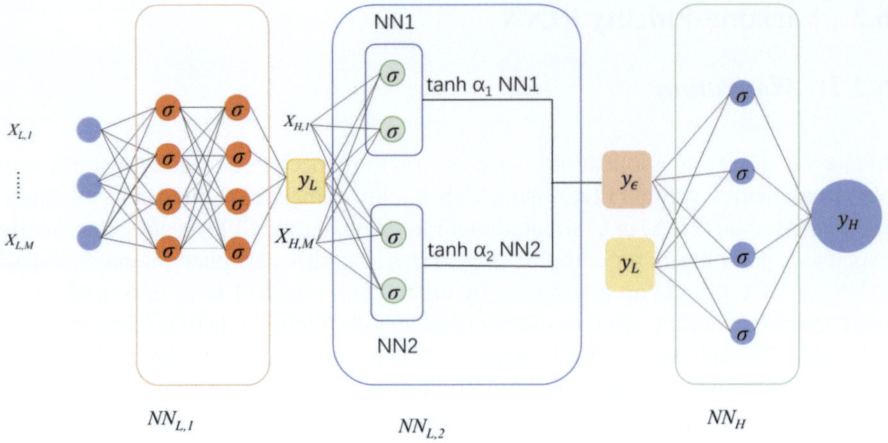

Fig. 6.2 Network structure of the proposed VF-PINN

where Y_H is the high-fidelity solution, Y_L is the low-fidelity solution, and $\delta(\mathbf{x})$ is a residual function approximated by the fusion network. Specifically, the fusion process is expressed as

$$Y_H = Y_L + \epsilon\,(\tanh(a_1 f_1(\mathbf{x}, Y_L)) + \tanh(a_2 f_2(\mathbf{x}, Y_L))), \tag{6.4}$$

where $f_1(\mathbf{x}, Y_L)$ is a linear function without an activation function, $f_2(\mathbf{x}, Y_L)$ is a nonlinear function, a_1 and a_2 are trainable parameters, and ϵ is a residual rate that scales the correction term. This formulation ensures the solution adheres to physical constraints enforced through the loss function.

The use of separate networks for low- and high-fidelity data allows the model to learn distinct features from each fidelity level, while the fusion network merges these features to produce a cohesive solution that balances efficiency and accuracy. This architecture is illustrated in Fig. 6.2, which depicts the parallel processing of multi-fidelity data and their subsequent fusion.

6.2.3 *Training Process*

The VF-PINN approximates the PDE solution as $Y = \hat{u}(\mathbf{x}; \theta)$, where θ represents the trainable parameters across all networks. The high-fidelity solution Y_H is expressed as a correction to the low-fidelity solution Y_L, formulated as

$$Y_H = Y_L + \delta(\mathbf{x}), \tag{6.5}$$

where $\delta(\mathbf{x})$ is the residual function approximated by the fusion network (NN_f). This residual term is specifically defined as

$$\delta(\mathbf{x}) = \epsilon\,(\tanh(a_1 f_1(\mathbf{x}, Y_L)) + \tanh(a_2 f_2(\mathbf{x}, Y_L))), \tag{6.6}$$

where $f_1(\mathbf{x}, Y_L)$ is a linear function without an activation function capturing linear adjustments to the low-fidelity solution, $f_2(\mathbf{x}, Y_L)$is a nonlinear function addressing more complex deviations. a_1, a_2 are trainable weights that modulate the contributions of the linear and nonlinear terms. ϵ is a scaling factor controlling the magnitude of the residual correction. tanh is the Hyperbolic tangent activation functions that introduce bounded nonlinearity for improved expressiveness and stability.

The training is guided by a composite loss function comprising three primary components:

1. **PDE Residual Loss** (MSE_r): This term enforces the satisfaction of the governing PDEs across the domain, using the fused output Y and its derivatives, computed via automatic differentiation, at collocation points $\{\mathbf{x}_i\}_{i=1}^{N_r}$:

$$\text{MSE}_r = \frac{1}{N_r}\sum_{i=1}^{N_r}\left|f\left(Y(\mathbf{x}_i), \frac{\partial Y}{\partial x_i}, \ldots, \frac{\partial^2 Y}{\partial x_i \partial x_j}\right)\right|^2, \tag{6.7}$$

   where $f(\cdot)$ is the PDE residual, and N_r collocation points are sampled using Latin hypercube sampling for uniform domain coverage.
2. **Boundary Condition Loss** (MSE_u): This component ensures compliance with specified boundary conditions at boundary points $\{\mathbf{x}_j\}_{j=1}^{N_u}$:

$$\text{MSE}_u = \frac{1}{N_u}\sum_{j=1}^{N_u}\left|Y(\mathbf{x}_j) - b_0(\mathbf{x}_j)\right|^2, \tag{6.8}$$

   where $b_0(\mathbf{x}_j)$ is the prescribed boundary condition, and N_u is the number of boundary points.
3. **Data Loss** (MSE_d): This term aligns the intermediate outputs Y_L and Y_H with the available low- and high-fidelity data:

$$\text{MSE}_d = \frac{1}{N_{HF}}\sum_{k=1}^{N_{HF}}|Y_H(\mathbf{x}_k) - u_{HF}(\mathbf{x}_k)|^2 + \lambda\frac{1}{N_{LF}}\sum_{m=1}^{N_{LF}}|Y_L(\mathbf{x}_m) - u_{LF}(\mathbf{x}_m)|^2, \tag{6.9}$$

   where N_{HF} and N_{LF} are the numbers of high- and low-fidelity data points, $u_{HF}(\mathbf{x}_k)$ and $u_{LF}(\mathbf{x}_m)$ are the corresponding observations, and $\lambda \in [0, 1]$ balances the influence of low-fidelity data to prioritize high-fidelity accuracy.

The total loss is a weighted sum:

$$\text{MSE} = \alpha_r \text{MSE}_r + \alpha_u \text{MSE}_u + \alpha_d \text{MSE}_d, \tag{6.10}$$

where α_r, α_u, and α_d are hyperparameters tuned to balance the contributions of physical constraints and data fidelity, critical for achieving an optimal solution.

The training algorithm for the VF-PINN framework, detailed in Algorithm 6.1, is designed to optimize the neural network parameters to approximate the solution of the given PDE by effectively integrating multi-fidelity data and enforcing physical constraints. The algorithm requires as input the low-fidelity data $\mathbf{x}_{LF}, u_{LF}$, high-fidelity data $\mathbf{x}_{HF}$, $\{\mathbf{x}_k, u_{HF}(\mathbf{x}_k)\}_{k=1}^{N_{HF}}$, the governing PDEs, boundary conditions, and a set of hyperparameters including weights for the loss terms $(\alpha_r, \alpha_u, \alpha_d)$, a balancing factor λ, the learning rate η, the maximum number of iterations N_{iter}, and the convergence tolerance ε. The parameters of the neural networks, specifically, the low-fidelity network (NN1), the high-fidelity network (NN2), and the fusion network (NN_f), are initialized, either randomly or using pre-trained values if available, and N_r collocation points and N_u boundary points are sampled using Latin hypercube sampling to ensure uniform coverage of the domain. The training process involves an iterative loop where, in each iteration, the intermediate outputs Y_L and Y_H are computed from the respective networks, fused into the predicted solution Y by NN_f, and used to evaluate the PDE residual loss MSE_r, the boundary condition loss MSE_u, and the data loss MSE_d. These losses are combined into a total loss $\text{MSE} = \alpha_r \text{MSE}_r + \alpha_u \text{MSE}_u + \alpha_d \text{MSE}_d$, which is minimized using the Adam optimizer with learning rate η. The loop continues until the total loss falls below the specified tolerance ε or the maximum number of iterations is reached, at which point the optimized neural network provides the predicted solution $Y(\mathbf{x})$.

Algorithm 6.1: VF-PINN Training Algorithm

Input: Low-fidelity data $\mathbf{x}_{LF}, u_{LF}$, high-fidelity data $\mathbf{x}_{HF}$, $\{\mathbf{x}_k, u_{HF}(\mathbf{x}_k)\}_{k=1}^{N_{HF}}$, PDEs, boundary conditions, hyperparameters $\alpha_r, \alpha_u, \alpha_d, \lambda$, learning rate η, iterations N_{iter}, tolerance ε.
Initialize parameters θ for NN1, NN2, and NN_f.
Sample collocation points $\{\mathbf{x}_i\}_{i=1}^{N_r}$ and boundary points $\{\mathbf{x}_j\}_{j=1}^{N_u}$ using Latin hypercube sampling.
for $t = 1$ to N_{iter} **do**
  Compute $Y_L = \text{NN1}(\mathbf{x}_{LF}; \theta_{\text{NN1}})$
  Compute $Y_H = \text{NN2}(\mathbf{x}_{HF}; \theta_{\text{NN2}})$
  Compute $Y = \text{NN\_f}(Y_L, Y_H; \theta_{\text{NN\_f}})$
  Compute derivatives via automatic differentiation.
  Evaluate MSE_r, MSE_u, and MSE_d.
  Compute total loss $\text{MSE} = \alpha_r \text{MSE}_r + \alpha_u \text{MSE}_u + \alpha_d \text{MSE}_d$.
  Update θ using Adam optimizer with learning rate η.
  if $\text{MSE} < \varepsilon$ **then**
    Break
  end if
end for
Output: Predicted solution $Y(\mathbf{x})$.

6.3 Experimental Results

6.3.1 Experimental Setup and Overview

In this section, two case studies are utilized to evaluate the efficacy of the VF-PINN framework: the Euler-Bernoulli beam and the two-dimensional Poisson Equation with variable source intensity. The objectives are to assess the framework's capability in assimilating multi-fidelity data to yield accurate solutions to PDEs, mitigate computational demands relative to high-fidelity simulations, and surpass the performance of standard PINNs under conditions of limited data availability. Performance is quantified using metrics such as the mean absolute percentage error (MAPE), root mean square error (RMSE), coefficient of determination (R^2), and computational training time. Each experiment is replicated five times to ensure statistical robustness, with results expressed as mean values accompanied by standard deviations.

6.3.2 Case Study 1: Euler-Bernoulli Beam with Variable Point Load

The Euler-Bernoulli beam equation governs the transverse deflection $w(x)$ of a slender beam subjected to a point load of magnitude q at position p, representing a foundational problem in structural mechanics [2, 3]. For a simply supported beam of length L, the governing equation is

$$EI\frac{d^4w}{dx^4} = q\delta(x - p), \quad 0 < x < L, \tag{6.11}$$

where E denotes Young's modulus, I the second moment of area, and δ the Dirac delta function. Boundary conditions are $w(0) = w(L) = 0$ and $\frac{d^2w}{dx^2}(0) = \frac{d^2w}{dx^2}(L) = 0$. To enable generalization, $q \in [0.5, 1.5]$ and $p \in [0.2, 0.8]$.

In this investigation, the parameters are set as $L = 1$ m and $EI = 1$ N·m^2. Low-fidelity (LF) data are generated using a finite element method (FEM) with 3 elements (4 nodes) for 100 random (q, p) configurations, yielding 1000 samples (10 points per configuration). High-fidelity (HF) data employ FEM with 30 elements (31 nodes) for 20 configurations, providing 200 samples (10 points per configuration). VF-PINN inputs consist of (x, q, p) from the respective fidelity levels.

The VF-PINN is trained for 8000 iterations using the Adam optimizer, followed by up to 1000 iterations of LBFGS optimization to achieve convergence. The Standard PINN is trained similarly with 8000 Adam iterations and up to 1000 LBFGS

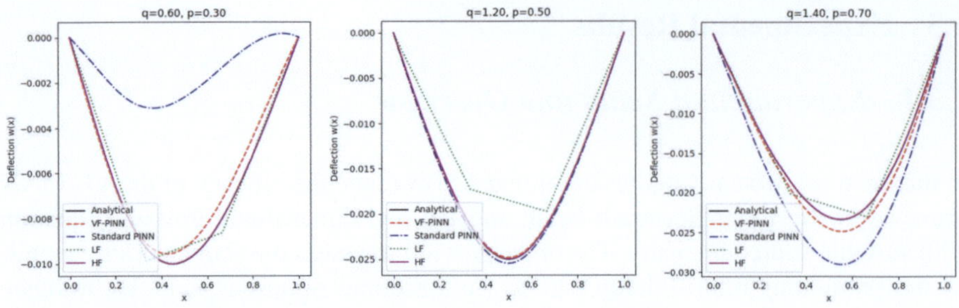

Fig. 6.3 Predicted deflection profiles for (q, p) relative to baselines

iterations. Figure 6.3 illustrates the predicted deflection profiles for three unseen configurations, $(q, p) \in \{(0.6, 0.3), (1.2, 0.5), (1.4, 0.7)\}$, comparing VF-PINN, Standard PINN, LF, and HF models against the analytical solution. The VF-PINN solution aligns closely with the analytical reference, exhibiting superior fidelity and generalization compared to the Standard PINN and coarse LF model.

To quantitatively assess the models' performance, three key metrics such as Root Mean Square Error (RMSE), Mean Absolute Percentage Error (MAPE), and coefficient of determination (R^2) are used. The evaluation is performed on a uniform 100-point test grid spanning the domain [0, 1], generated using $\mathbf{x}_{\text{test}} = \text{linspace}(0, 1, 100)$. For each unseen configuration $(q, p) \in \{(0.6, 0.3), (1.2, 0.5), (1.4, 0.7)\}$ (selected to cover low, medium, and high load magnitudes and positions within the training ranges [0.5, 1.5] for q and [0.2, 0.8] for p, but not exactly matching any training (q, p) pairs), the models predict $w(x, q, p)$ on this grid. The LF and HF baselines are obtained by solving the FEM equations on their respective meshes and interpolating to the test grid using linear interpolation, with boundary values enforced to zero for consistency.

The data for LF and HF originate from numerical solutions of the beam equation via a finite element solver, which assembles stiffness matrices for Hermite beam elements and solves the resulting linear system with simply supported boundary conditions. Metrics are computed for each configuration and averaged across the three (q, p) pairs to yield mean performance. The results, including training times, are summarized in Table 6.1. It can be observed that the VF-PINN achieves the lowest errors, demonstrating its effectiveness in leveraging multi-fidelity data for improved accuracy and generalization.

6.3.3 *Case Study 2: Two-Dimensional Poisson Equation with Variable Source Intensity*

The two-dimensional Poisson equation models phenomena such as electrostatic potential or steady-state heat conduction in a unit square domain $\Omega = [0, 1]^2$ [4].

Table 6.1 Performance metrics for the Euler-Bernoulli beam case across three unseen configurations

Metric	($q = 0.6, p = 0.3$)			($q = 1.2, p = 0.5$)			($q = 1.4, p = 0.7$)		
	HF	Standard PINN	VF-PINN	HF	Standard PINN	VF-PINN	HF	Standard PINN	VF-PINN
MAPE (%)	0.1	8.4	3.8	0.2	8.8	4.0	0.3	9.2	4.2
RMSE	0.0004	0.002	0.006	0.0005	0.005	0.001	0.0006	0.012	0.003
R^2	0.999	0.885	0.977	0.999	0.903	0.976	0.999	0.851	0.955

Its versatility makes it a critical model for applications ranging from semiconductor design to gravitational field analysis. The governing equation with a sinusoidal source term is

$$-\Delta u = q \cdot 2(k\pi)^2 \sin(k\pi x)\sin(k\pi y), \quad (x, y) \in \Omega, \tag{6.12}$$

where $k = 2.2$, $\Delta u = \frac{\partial^2 u}{\partial x^2} + \frac{\partial^2 u}{\partial y^2}$ and q represents the source intensity. Dirichlet boundary conditions are imposed: $u = 0$ on $\partial\Omega$. To facilitate generalization, $q \in [0.5, 1.5]$.The sinusoidal source term with wave number $k = 2.2$ introduces a higher spatial frequency compared to the standard $k = 1$, creating a more complex field that challenges the numerical methods' ability to capture rapid variations. The variable source intensity q allows the model to generalize across different load magnitudes, simulating realistic scenarios where source strength varies, such as in heat transfer with variable heat sources or electrostatics with fluctuating charges.

In this investigation, low-fidelity (LF) data are generated using a finite difference method on a coarse grid ($N = 20$) for 60 random q configurations, yielding 1200 samples (20 points per configuration). High-fidelity (HF) data utilize a fine grid ($N = 300$) for 5 configurations, providing 50 samples (10 points per configuration). VF-PINN inputs consist of (x, q, p) from the respective fidelity levels.

The VF-PINN and Standard PINN are trained for 10,000 iterations using the Adam optimizer, followed by up to 1000 iterations of LBFGS optimization to achieve convergence. Figure 6.5 illustrates the predicted two-dimensional solution fields for $q \in \{0.6, 1.2, 1.4\}$, comparing VF-PINN, Standard PINN, LF, and HF models against the analytical solution. One-dimensional slices at $y = 0.5$ are shown in Fig. 6.4, highlighting the models' performance along a cross-section of the domain. The VF-PINN solution aligns closely with the analytical reference, exhibiting superior fidelity and generalization compared to the Standard PINN and coarse LF model.

To quantitatively assess the performance of the models, we evaluate several key metrics, including the Root Mean Square Error (RMSE), Mean Absolute Percentage Error (MAPE), and Coefficient of Determination (R^2), on a 50×50 test grid across three unseen configurations ($q = 0.6, 1.2, 1.4$) against the analytical solution. For each configuration q, the models are queried to produce predictions u_{pred} at these grid points, while the true values u_{true} are computed from the analytical solution.

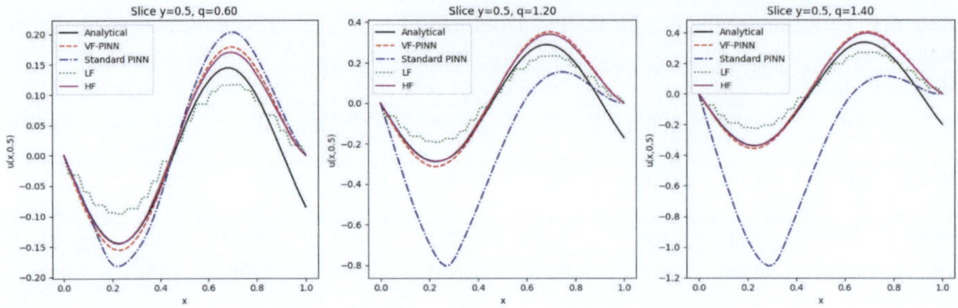

Fig. 6.4 Predicted solution fields

Table 6.2 Performance metrics for the two-dimensional Poisson equation case

Metric	$(q = 0.6)$			$(q = 1.2)$			$(q = 1.4)$		
	HF	Standard PINN	VF-PINN	HF	Standard PINN	VF-PINN	HF	Standard PINN	VF-PINN
MAPE (%)	1.3	7.3	5.6	2.2	7.8	5.0	1.8	6.2	4.2
RMSE	0.058	0.124	0.073	0.076	0.139	0.098	0.042	0.081	0.061
R^2	0.916	0.785	0.862	0.929	0.803	0.856	0.908	0.824	0.859

The LF and HF baseline results are obtained by interpolating the finite difference solutions (coarse grid for LF, fine grid for HF) onto the 50×50 test grid using nearest-neighbor interpolation, with boundary values enforced to zero for consistency. The predictions are flattened into vectors for metric computation. The averaged results across the three configurations are summarized in Table 6.2, demonstrating that the VF-PINN solution aligns closely with the analytical reference, exhibiting superior fidelity and generalization compared to the Standard PINN and coarse LF model.

6.4 Summary

This chapter introduces a generalized VF-PINN framework that assimilates low- and high-fidelity data through a multi-network architecture to furnish efficient and accurate solutions to PDEs, with generalization to varying parameters. Case studies encompassing the Euler-Bernoulli beam with variable point load and the two-dimensional Poisson equation models, with fidelity levels delineated by grid resolution, affirm the framework's versatility across engineering domains. The comparative analysis highlights the VF-PINN's capacity to advance design optimization by harmonizing computational exigency and predictive fidelity, thereby establishing it as a potent instrument for variable-fidelity surrogate modeling.

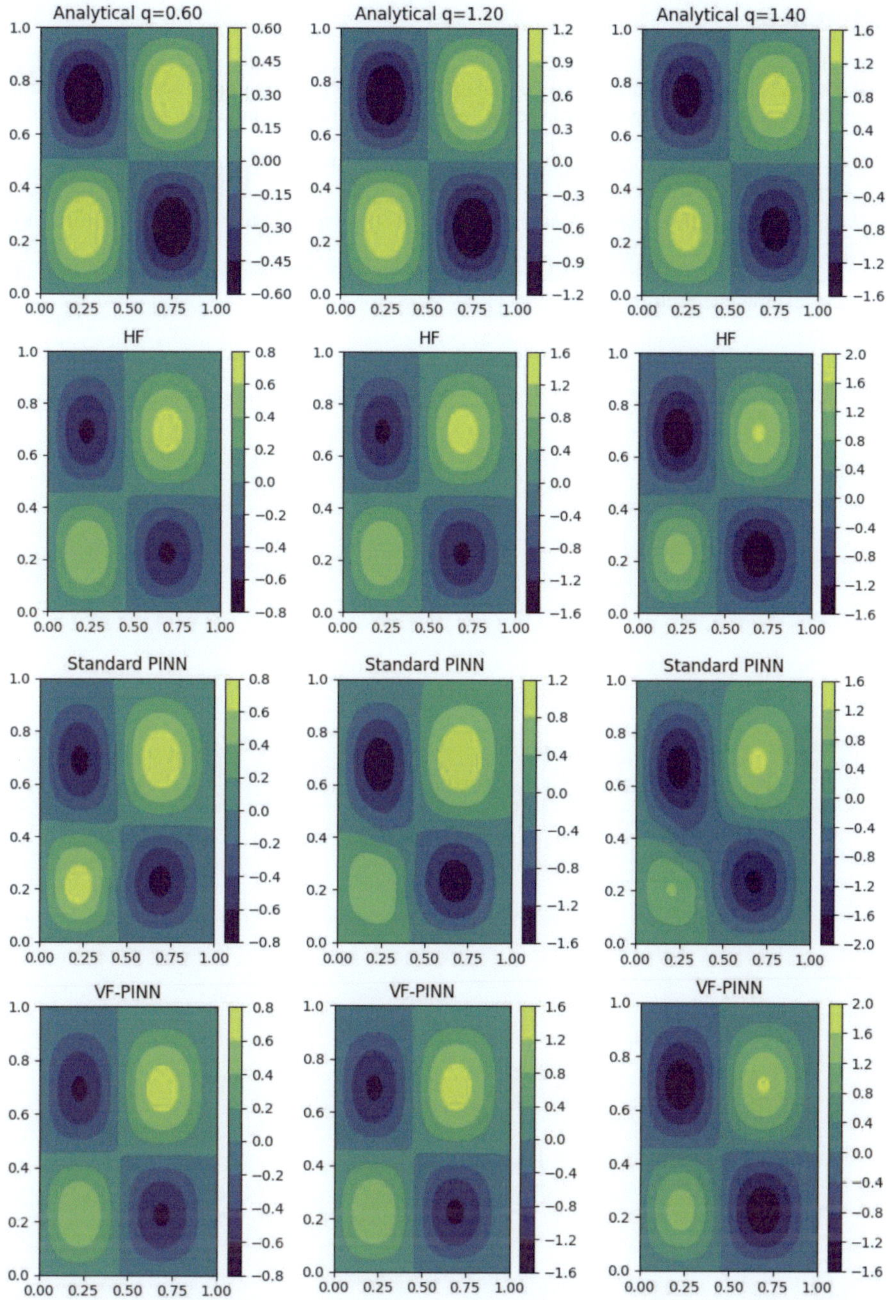

Fig. 6.5 Contour plots of analytical and predicted solutions for three unseen configurations ($q = 0.6, 1.2, 1.4$)

References

1. Raissi M, Perdikaris P, Karniadakis GE (2019) Physics-informed neural networks: a deep learning framework for solving forward and inverse problems involving nonlinear partial differential equations. J Comput Phys 378:686–707
2. Bauchau OA, Craig JI (2009) Euler-Bernoulli beam theory. In: Structural analysis. Springer, pp 173–221
3. Haider JA, Zaman FD, Lone SA et al (2023) Exact solutions of Euler–Bernoulli beams. Mod Phys Lett B 37(33):2350161
4. Hackbusch W (2017) The Poisson equation. In: Elliptic differential equations: theory and numerical treatment. Springer, pp 29–42

Chapter 7
Multi-fidelity Transfer Learning Model Based on Dynamic Task-Weighted Loss

Recent advancements in deep learning (DL) have significantly advanced the solution of complex engineering problems. Central to DL, deep neural networks (DNNs) extend traditional multilayer perceptrons by incorporating hierarchical feature learning, which extracts and transforms input features prior to model training. This mechanism enables superior representation of complex patterns and enhances the ability to handle high-dimensional data, providing improved generalization performance relative to conventional surrogate models. Consequently, DNNs are increasingly replacing traditional machine learning approaches in engineering optimization.

To substantially reduce computational costs in optimization, variable-fidelity surrogate modeling techniques (VFSM), which combines high-fidelity (HF) and low-fidelity (LF) models have emerged as an effective solution. These methods effectively address the trade-off between surrogate model predictive accuracy and construction expenses. Nevertheless, existing VFSM approaches primarily rely on predefined relational assumptions, which may not fully capture the intricate relationships between HF and LF models. To overcome this limitation, this chapter proposes a novel multi-fidelity surrogate modeling framework based on deep transfer learning. By utilizing deep transfer learning, the proposed method identifies shared latent features between HF and LF models. Comparative experiments across different sample sizes with established MFSM techniques demonstrate the enhanced performance of this deep transfer learning-based multi-fidelity surrogate modeling approach.

7.1 Construction of Multi-fidelity Transfer Learning Model Based on Dynamic Task-Weighted Loss

Multi-fidelity surrogate modeling based on linear assumptions offers simplicity in modeling, but is constrained by limited accuracy. On the other hand, nonlinear assumption-based MF surrogates can capture more complex relationships between models of different fidelities. However, they are prone to overfitting due to numerous

J. Yi et al., *Variable-Fidelity Surrogate*, Engineering Applications of Computational Methods 25, https://doi.org/10.1007/978-981-95-5527-7_7

hyperparameters and scarce HF samples, resulting in low precision. Therefore, to construct an accurate MF surrogate, it is imperative to address key challenges such as the intrinsic relationship between models and the scarcity of HF samples. To this end, this chapter introduces a novel multi-fidelity surrogate modeling method termed Harmonic Domain Regressor with Dynamic Task-Weighted (HDR-DTW) strategy. This approach utilizes the robust training capacity of deep transfer learning to elucidate the relationship between HF and LF models and employs a dynamic task-weighted loss strategy to accelerate convergence and enhance predictive accuracy.

7.1.1 MF Surrogate Model Inspired by Transfer Learning

Transfer learning (TL) facilitates the application of common features or general principles learned from different tasks to new tasks, encompassing feature-based, instance-based, and parameter-based transfer learning. It has found extensive applications in object detection, image recognition, mechanical fault diagnosis, etc. TL involves source domain samples $D_s = \{x_s, P_s(x)\}$, source task T_s, target domain samples $D_t = \{x_t, P_t(x)\}$, and target task T_t, where $P(\cdot)$ denotes the distribution space of the samples. TL aims to utilize the information from D_s and T_s to efficiently train D_t in T_t. Generally, D_s represents a domain with abundant labeled samples containing comprehensive task information, while D_t is a domain with few or no labeled samples. When $P_s(x_s) = P_t(x_t)$, it is referred to as homogeneous transfer learning. In contrast, heterogeneous transfer learning occurs when $P_s(x_s) \neq P_t(x_t)$. The core mechanism of TL lies in exploiting shared attributes between the source and target domains. Successful TL transfers knowledge from the source to the target domain. However, it is crucial to recognize that such transfer is not always beneficial. In the absence of common features between domains, TL may lead to negative transfer. Compared to previous machine learning approaches, TL offers two primary advantages: (1) addressing small sample size issues, where TL can train a new task model with relatively few labeled data based on similar models trained on large datasets, and (2) reducing training costs, as TL obviates the need for complete training cycles on the entire dataset required for new model training.

By comparing the main ideas of the TL and MF surrogate modeling, as illustrated in Fig. 7.1, several similarities can be observed. In TL, the source task and target task models correspond to the LF and HF models in the MF surrogate modeling. The process of fine-tuning a pretrained model to obtain a target task model bears resemblance to adjusting the LF model via a scaling function to acquire the HF model. Given the highly complex nonlinear relationship between LF and HF models, accurately describing this relationship and constructing a precise surrogate model based on limited high-fidelity samples and assumptions pose significant challenges. The advanced nonlinear approximation capabilities of DNNs can be harnessed to address these challenges in MF surrogate modeling through deep transfer learning (DTL).

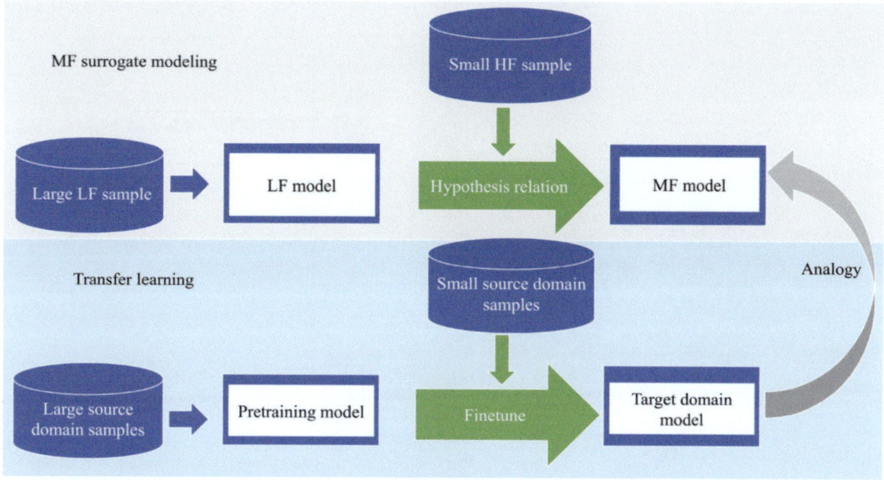

Fig. 7.1 Comparison between MF surrogate model and TL

7.1.2 Harmonic Domain Regressor

Accurately extracting common features from the source and target domains is pivotal in TL. Among the prevalent TL approaches, model fine-tuning requires identifying appropriate weight-sharing layers to prevent negative transfer. To circumvent this issue, a novel Harmonic Domain Regressor (HDR) is proposed, which integrates samples from both source and target domains, thereby eliminating the need for weight-sharing layer selection. The HDR primarily employs three networks: a feature extraction network G_f and two task-fitting networks (G_{r1} and G_{r2}), as depicted in Fig. 7.2. The feature extraction network G_f is tasked with extracting common features from both domains, while G_{r1} and G_{r2} fit the source and target tasks, respectively. Consequently, the source domain model (LF model) comprises G_f and G_{r1}, and the target domain model (HF model) consists of G_f and G_{r2}. The output of the feature extraction network serves as input to the subsequent two neural networks, with all three networks being trained collaboratively using samples from both source and target domains.

7.1.3 Dynamic Task-Weighted Loss

Although joint training of source and target domain samples in neural networks circumvents weight-sharing layer selection, it effectively enables parameter sharing and separation between tasks while providing independent spaces to prevent interference. However, during training, the following issues may arise due to task differences, such as disparities in sample quantities:

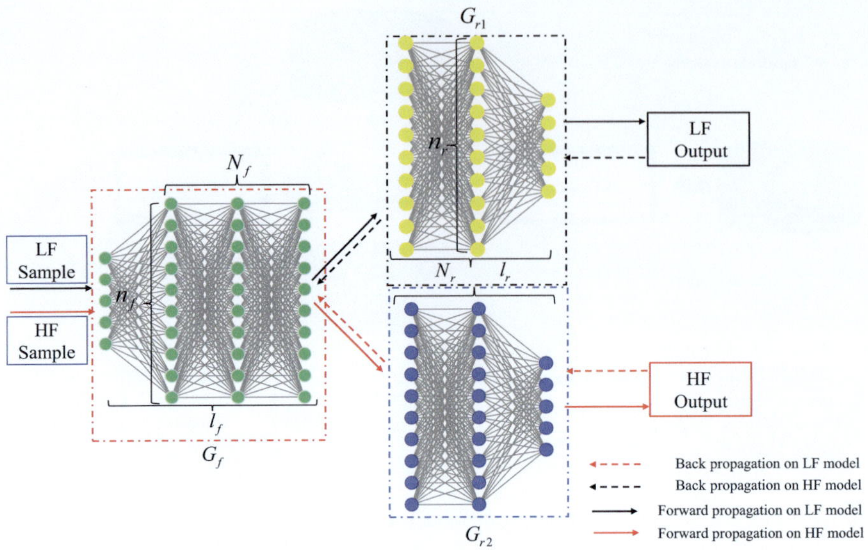

Fig. 7.2 Network architecture of HDR

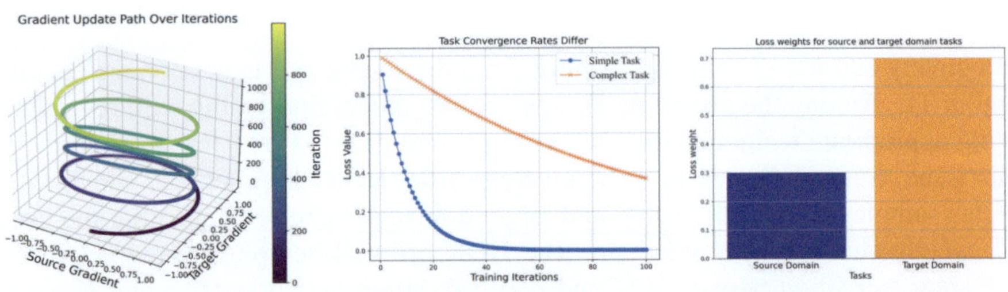

(a) Divergent gradients (b) Varied convergence rates (c) Different loss weights

Fig. 7.3 Issues arising from joint training

(1) Divergent gradient update directions: Within the same parameter set, the gradient update directions for different tasks may diverge, causing parameter fluctuations and potentially leading to negative transfer effects between tasks, as illustrated in Fig. 7.3a.
(2) Varied convergence rates: Simple tasks tend to converge faster, while complex tasks converge more slowly. After several training iterations, this disparity may result in overfitting of simple tasks and underfitting of complex tasks, as depicted in Fig. 7.3b.
(3) Impact of task-specific loss weights on training dynamics: Tasks with higher loss weights may dominate the parameter update process, potentially overshadowing the learning objectives of tasks with lower weights, as shown in Fig. 7.3c.

Considering the abundance of source domain samples, it is essential to assign different weight losses to source and target domain tasks during joint training to ensure that the source domain task does not dominate parameter updates. Moreover, the model is expected to exhibit superior predictive performance on the target task post-training. Consequently, the training process is divided into two distinct phases: feature extraction and regression. During the feature extraction phase, the focus is on extracting commonalities between tasks, with the source domain task designated as the primary task and the target domain task as the auxiliary task. The total loss L can be expressed as

$$L = \alpha L_1 + \beta L_2, \tag{7.1}$$

$$\alpha = 1, \beta = L_1/(L_1 + L_2). \tag{7.2}$$

Here, L_1 represents the source domain loss, and L_2 denotes the target domain loss. As feature extraction progresses and L_1 diminishes, the value of β decreases correspondingly, thus reducing the influence of the target task (auxiliary task) on feature extraction and enabling the source domain task (primary task) to extract common features more effectively. In the regression phase, the emphasis shifts to achieving excellent performance on the target task, with the source domain task serving as the auxiliary task and the target domain task as the primary task. Thus, the total loss L in this phase is represented as

$$L = \alpha L_1 + \beta L_2, \tag{7.3}$$

$$\alpha = L_2/(L_1 + L_2), \beta = 1. \tag{7.4}$$

It is evident that due to the larger weight of the target task loss, the network updates tend to favor the target task. As L_2 decreases, α diminishes, reducing the influence of the source domain task (auxiliary task) and ensuring superior performance of the network on the target task (primary task).

7.1.4 Main Steps

Based on the aforementioned ideas, this section proposes a multi-fidelity surrogate modeling method combining Harmonic Domain Regression and Deep Transfer Learning with Dynamic Task-Weighted Loss (HDR-DTW). The framework is illustrated in Fig. 7.4.

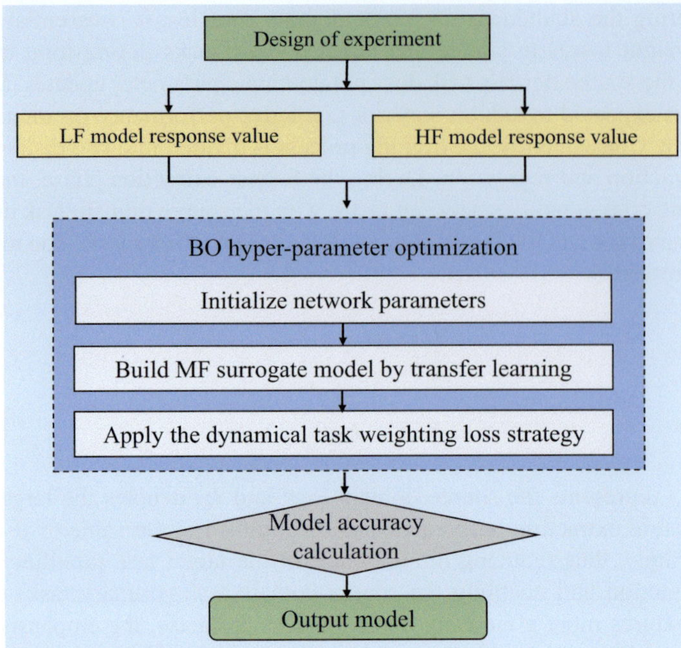

Fig. 7.4 Flowchart of HDR-DTW

The specific steps are as follows:

(1) Experimental design. Latin Hypercube Sampling (LHS) [1] is employed, a widely used method in experimental design. This multidimensional stratified sampling technique divides the design variable space into uniform subregions and selects one sample point in each subregion, ensuring uniform distribution of sample points and facilitating the generation of an arbitrary number of samples. Based on the cost ratio r_c of LF and HF models, the number of LF and HF sample points is designated as n_s and n_t, respectively. Unlike previous nested experimental designs [2], there is no requirement for HF samples to be nested within LF samples.
(2) Response computation. The simulation responses of n_s LF sample points and n_t HF sample points are calculated, serving as the source and target domain samples for transfer learning, respectively.
(3) Based on n_s source domain and n_t target domain sample points, the feature extraction network G_f and networks G_{r1} and G_{r2} are initialized. For a fully connected DNN with n hidden layers, the primary components of its model parameters are weights w and biases b. The output of the i-th layer in the DNN is given by

$$z^{(i)} = f(w^{(i-1)}z^{(i-1)} + b(i)) \qquad \forall i \in \{1, 2, \ldots, n\}. \tag{7.5}$$

Here, $z^{(i)}$ denotes the output of layer i, $z^{(i-1)}$ represents the output of layer $i-1$, and f signifies the activation function. Various activation functions exist, including the logistic (Sigmoid) function, hyperbolic tangent (Tanh) function, and rectified linear unit (ReLU) function. The weights w and biases b can collectively be represented by the parameter $\boldsymbol{\theta}$ ($\boldsymbol{\theta} = \{w^{(i)}, b(i)\}_{i=1}^{n+1}$). $\boldsymbol{\theta}$ can be optimized by minimizing the following formula:

$$L(\boldsymbol{\theta}) = \frac{1}{N}\sum_{i=1}^{N}(y_i - \tilde{f}(x_i; \boldsymbol{\theta}))^2. \tag{7.6}$$

Here, $L(\boldsymbol{\theta})$ represents the mean squared error (MSE) loss, (x_i, y_i) denotes training data, and N is the number of samples. $\tilde{f}(x_i; \boldsymbol{\theta})$ indicates the network's prediction. To minimize the loss function, the widely used Adam algorithm is employed, with the learning rate r_l in the Adam algorithm serving as a hyperparameter for subsequent optimization. The initial training process's total loss is given by

$$\tilde{f_1}(x_i; \boldsymbol{\theta}_f; \boldsymbol{\theta}_{r1}) = G_{r1}(G_f(x_i; \boldsymbol{\theta}_f); \boldsymbol{\theta}_{r1}), \tag{7.7}$$

$$\tilde{f_2}(x_i; \boldsymbol{\theta}_f; \boldsymbol{\theta}_{r2}) = G_{r2}(G_f(x_i; \boldsymbol{\theta}_f); \boldsymbol{\theta}_{r2}), \tag{7.8}$$

$$L_1(\boldsymbol{\theta}_f; \boldsymbol{\theta}_{r2}) = \frac{1}{n_s}\sum_{i=1}^{n_s}(y_i - \tilde{f_1}(x_i; \boldsymbol{\theta}_f; \boldsymbol{\theta}_{r1}))^2, \tag{7.9}$$

$$L_2(\boldsymbol{\theta}_f; \boldsymbol{\theta}_{r2}) = \frac{1}{n_t}\sum_{i=1}^{n_t}(y_i - \tilde{f_2}(x_i; \boldsymbol{\theta}_f; \boldsymbol{\theta}_{r2}))^2, \tag{7.10}$$

$$L = \alpha L_1 + \beta L_2, \tag{7.11}$$

where α and β represent hyperparameters that control task inclination. Assume the total number of training iterations is T, when the current iteration $t \leq 0.5T$, $\alpha = 1$, and $\beta = \frac{L_1}{L_1+L_2}$ are set. When $0.5T < t \leq T$, $\alpha = \frac{L_2}{L_1+L_2}$ and $\beta = 1$ are set. The Adam algorithm is utilized to optimize the loss function L, thereby obtaining the weights and thresholds of the feature extraction network G_f, G_{r1}, and G_{r2}.

(4) Hyperparameter optimization. Given the significant impact of neural networks on their hyperparameters, the Bayesian Optimization (BO) method is employed to search for optimal hyperparameters. The objective function $f_H(x)$ is defined by minimizing the validation error. The hyperparameter optimization variables include some structural parameters of the network and the learning rate. In this

study, the two regressors utilize identical network structures. Consequently, the variables for BO encompass the number of hidden layers N_f in the feature extraction network, the number of neurons n_f in the hidden layers, the number of hidden layers N_r in the regression network, the number of neurons n_r in the hidden layers, and the gradient threshold g for gradient clipping to prevent gradient explosion. Gradient explosion refers to a phenomenon in deep learning training in which gradients accumulate excessively large values, potentially causing significant weight updates and destabilizing training. To ensure efficient gradient updates, learning rates $l_f = 0.0001$ and $l_r = 0.0001$ are set. By adjusting hyperparameters through BO, the most suitable parameters for the model are selected, yielding a multi-fidelity surrogate model.

7.1.5 One-Dimensional Function Example

This section elucidates the proposed HDR-DTW modeling method through a one-dimensional function. The expressions for the HF and LF models are as follows:

$$f_H(x) = (6x - 2)^2 \sin(12x - 4), \tag{7.12}$$

$$f_L(x) = 0.5 f_H(x) + 10(x - 0.5) - 5. \tag{7.13}$$

Here, $f_H(x)$ and $f_L(x)$ represent the responses of the HF and LF models, respectively, as shown in Fig. 7.5a. The HF and LF models exhibit consistent overall trends. Optimal LHS is employed to sample the LF and HF models, yielding 20 LF samples and 6 HF samples, which serve as the source and target data, respectively. Initially, a single-fidelity Kriging model is used to fit the HF data, resulting in the single-fidelity Kriging model depicted in Fig. 7.5b. Due to the limited number of high-fidelity samples, the surrogate model's accuracy is low, failing to fully capture the trend of the HF model. Additionally, the traditional multi-fidelity surrogate modeling method, Kriging-MFAM, is illustrated in Fig. 7.5c. It can be observed that with the aid of LF samples, the surrogate model demonstrates improved proximity to the HF model compared to the single-fidelity model based solely on HF samples. This validates the ability of multi-fidelity surrogate modeling techniques to leverage LF model information to enhance predictive accuracy. However, since Kriging-MFAM relies on prior relationship assumptions, it struggles to fully characterize the complex relationship between HF and LF models, resulting in significant discrepancies between Kriging-MFAM and the HF model. Figure 7.5d displays the predictive outcomes of the introduced HDR-DTW method. By jointly training the feature extractor $(N_f = 2, n_f = 2, l_f = 1e - 4)$ and $(N_r = 2, n_r = 2, lr = 1e - 4)$ using the proposed dynamic task-weighted method, HDR-DTW exhibits a closer alignment with the true HF model. This indicates that the HDR-DTW-MFAM method proposed in this study can accurately extract common features between HF and LF models

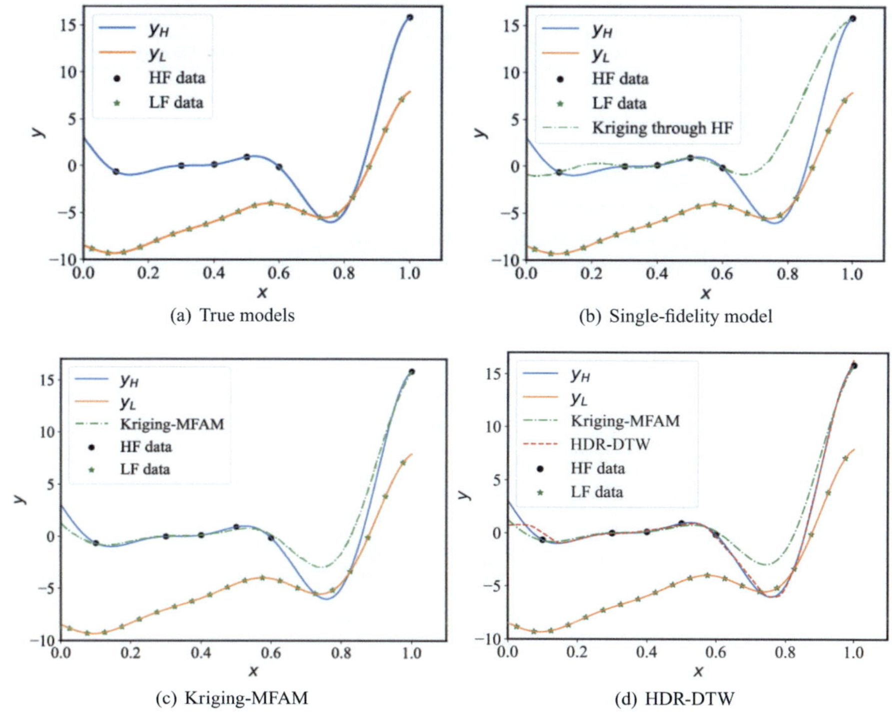

(a) True models (b) Single-fidelity model
(c) Kriging-MFAM (d) HDR-DTW

Fig. 7.5 Different surrogate models for 1D function

through deep transfer learning and achieve a multi-fidelity surrogate based on transfer learning with superior predictive accuracy via effective training.

7.2 Numerical Results

To verify the efficacy and applicability of the proposed surrogate modeling method, this section evaluates its performance across 10 benchmark functions of varying dimensions. Furthermore, the proposed method (HDR-DTW) is compared to four widely used MF surrogate modeling techniques. The impact of the dynamic task-weighting strategy is also analyzed in detail.

7.2.1 *Experimental Setup*

To comprehensively compare different multi-fidelity surrogate modeling (MFAM) methods, two common prediction accuracy metrics are selected: the normalized root

Table 7.1 Test functions

Example	Dimension	Expression	Variable range
F1	2	$f_H(\boldsymbol{x}) = 100(x_1^2 - x_2) + (x_1 - 1)^2$ $f_L(\boldsymbol{x}) = f_H(0.75x_1, x_2) - x_1 + 2\sin(x_1) + 2x_2$	$[-2, 2]^2$
F2	2	$f_H(\boldsymbol{x}) = 100(x_1^2 - x_2) + (x_1 - 1)^2$ $f_L(\boldsymbol{x}) = 70(x_1^2 - x_2)^2 + 0.7(x_1 - 1)^2$	$[-2.0, 2.0]^2$
F3	2	$f_H(\boldsymbol{x}) = (x_1^2 + x_2 - 11)^2 + (x_2^2 + x_1 - 7)^2$ $f_L(\boldsymbol{x}) = f_H(0.5x_1, 0.8x_2) + x_2^3 - (x_1 + 1)^2$	$[-3, 3]^2$
F4	3	$f_H(\boldsymbol{x}) = (x_1 - 1)^2 + (x_1 - x_2)^2 + x_1x_2 + 0.5$ $f_L(\boldsymbol{x}) = 0.2f_H(\boldsymbol{x}) - 0.5x_1 - 0.2x_1x_2 - 0.1$	$[0, 1]^3$
F5	4	$f_H(\boldsymbol{x}) = \frac{2}{3}e^{x_1 - x_2} - x_4 \sin x_3 + x_3$ $f_L(\boldsymbol{x}) = 1.2f_H(\boldsymbol{x}) - 1$	$[0, 1]^4$
F6	6	$f_H(\boldsymbol{x}) = \sum_{i=1}^{6} x_i^2 + \left(\sum_{i=1}^{6} 0.5ix_i\right)^2 + \left(\sum_{i=1}^{6} 0.5ix_i\right)^4$ $f_L(\boldsymbol{x}) = 2.5\left[\left(\sum_{i=1}^{6} 0.5ix_i\right)^2 + \left(\sum_{i=1}^{6} 0.5ix_i\right)^4\right]$	$[-5, 10]^6$
F7	6	$f_H(\boldsymbol{x}) = \sum_{i=1}^{5}\left[100(x_i^2 - x_{i+1})^2 + (x_i - 1)^2\right]$ $f_L(\boldsymbol{x}) = \sum_{i=1}^{5}\left[100(x_i^2 - 4x_{i+1})^2 + (x_i - 1)^2\right]$	$[0, 1]^6$
F8	8	$f_H(\boldsymbol{x}) = \sum_{i=1}^{2}\left[(x_{4i-3} + 10x_{4i-2})^2 + 5(x_{4i-1} - x_{4i})^2 + (x_{4i-2} - 2x_{4i-1})^4 + 10(x_{4i-3} + 10x_{4i})^4\right]$ $f_L(\boldsymbol{x}) = \sum_{i=1}^{2}\left[(x_{4i-3} + 10x_{4i-2})^2 + 125(x_{4i-1} - x_{4i})^2 + (x_{4i-2} - 2x_{4i-1})^4 + 10(x_{4i-3} + 10x_{4i})^4\right]$	$[0, 1]^8$
F9	16	$f_H(\boldsymbol{x}) = (x_1 - 1)^2 + \sum_{i=2}^{16} i(2x_i^2 - x_{i-1})^2$ $f_L(\boldsymbol{x}) = 0.9f_H(\boldsymbol{x}) + 0.2\sum_{i=1}^{15} x_ix_{i+1} - 300$	$[-5, 5]^{16}$
F10	20	$f_H(\boldsymbol{x}) = (x_1 - 1)^2 + \sum_{i=2}^{20}(2x_i^2 - x_{i-1})^2$ $f_L(\boldsymbol{x}) = 0.8f_H(\boldsymbol{x}) + 0.2\sum_{i=2}^{20} 0.4x_{i-1}x_i - 50$	$[-3, 3]^{20}$

mean square error (NRMSE) and the R^2 coefficient of determination. The performance of the proposed multi-fidelity surrogate modeling method is evaluated using 10 test functions of varying dimensions, detailed in Table 7.1, sourced from prior literature [3]. The four comparative MF surrogate modeling methods include (1) MF surrogate modeling based on linear relationship assumptions; (2) MF surrogate modeling based on nonlinear relationship assumptions, such as multi-fidelity neural

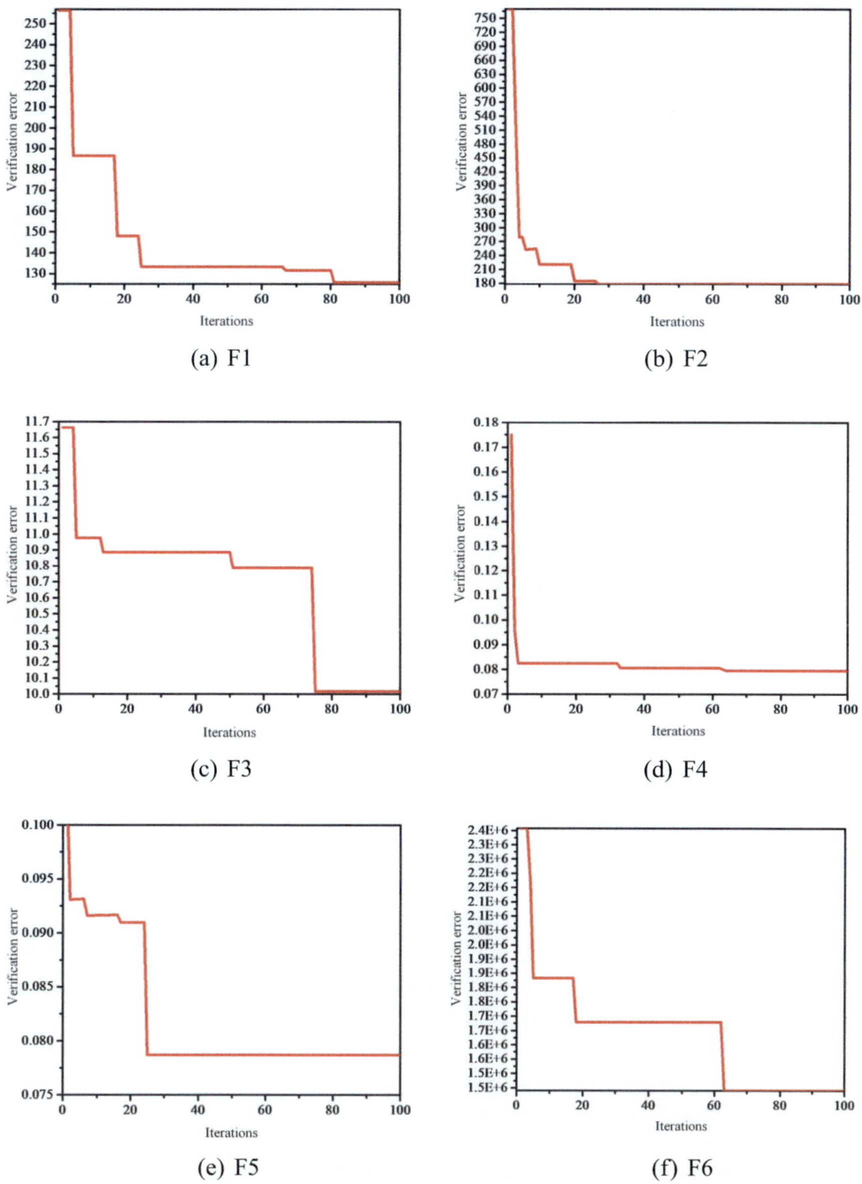

Fig. 7.6 Convergence process of hyperparameter optimization based on BO (F1–F6)

networks (DNN-MFAM) [3]; and (3) fine-tuning-based transfer learning methods (FT-MFAM), which primarily involve training an LF neural network with LF samples, fixing the weights of the initial layers of the LF neural network, and then fine-tuning the last layer using HF samples.

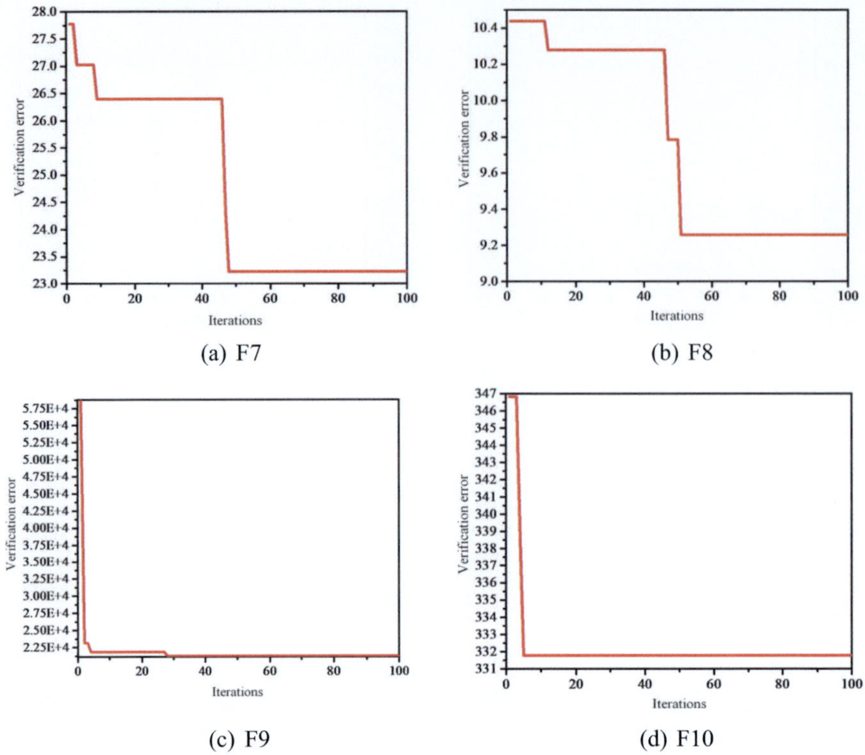

Fig. 7.7 Convergence process of hyperparameter optimization based on BO (F7–F10)

To ensure a fair comparison of the predictive performance of various methods, all models should utilize the same space-filling sampling approach. Consequently, optimal Latin hypercube sampling is employed for both HF and LF samples. Different combinations of HF and LF sample points can influence the prediction accuracy of MF surrogate models. Additionally, a sufficient number of LF sample points is necessary to establish a reliable surrogate model. Insufficient LF samples may lead to low prediction accuracy of the constructed MF surrogate model. Hence, this study selects an adequate number of LF samples to investigate the impact of varying HF sample quantities on model accuracy.

The number of LF samples is set to 50n, and the number of HF samples to 5n and 10n, where n represents the dimension of design variables. A multi-fidelity transfer learning model based on deep transfer learning with dynamic weighted loss (HDR-DTW) was constructed for each test case, and the network hyperparameters were optimized using the BO method. For each example, the number of iterations was set to 100. Figures 7.6 and 7.7 illustrate the optimization iteration process during Bayesian optimization. It can be observed that the objective function converges rapidly. Table 7.2 presents the optimal network hyperparameters obtained for each

Table 7.2 Variable ranges and optimized network hyperparameters for all test problems

Variable range						Optimized result
Example	N_f	N_r	n_f	n_r	g	$[N_f, N_r, n_f, n_r, g]$
F1	[20, 300]	[20, 300]	[2, 5]	[2, 5]	[0.2, 5]	[119, 38, 2, 6, 3.611]
F2	[20, 300]	[20, 300]	[2, 5]	[2, 5]	[0.2, 5]	[178, 53, 4, 2, 2.251]
F3	[20, 300]	[20, 300]	[2, 5]	[2, 5]	[0.2, 5]	[98, 132, 4, 2, 2.536]
F4	[20, 300]	[20, 300]	[2, 5]	[2, 5]	[0.2, 5]	[65, 72, 2, 2, 2.202]
F5	[20, 300]	[20, 300]	[2, 5]	[2, 5]	[0.2, 5]	[282, 39, 9, 8, 3.006]
F6	[20, 300]	[20, 300]	[2, 9]	[2, 9]	[0.2, 5]	[100, 162, 8, 8, 8.407]
F7	[20, 300]	[20, 300]	[2, 9]	[2, 9]	[0.2, 5]	[231, 184, 3, 2, 1.985]
F8	[20, 300]	[20, 300]	[2, 9]	[2, 9]	[0.2, 5]	[229, 28, 3, 5, 2.829]
F9	[20, 300]	[20, 300]	[2, 9]	[2, 9]	[0.2, 5]	[233, 66, 7, 7, 4.536]
F10	[20, 300]	[20, 300]	[2, 9]	[2, 9]	[0.2, 5]	[201, 31, 3, 9, 2.646]

test function when constructing the HDR-DTW model. The accuracy of the HDR-DTW model is significantly dependent on the selection of network hyperparameters. Therefore, during the model training process, the weights of the feature extraction stage and the regression stage are equal. The total number of training iterations is set to 1000, with each stage accounting for 500 iterations. Consequently, this study does not investigate the proportion of weights between the feature extraction stage and the regression stage during model iterative training.

7.2.2 Comparison Results

To mitigate random errors from single runs, each MF surrogate modeling method was repeated 10 times for each test case with the same samples. Each time, 500 test points were randomly selected, and the accuracy indices of the constructed surrogate models were calculated. The average values of NRMSE and R^2 over 10 runs are presented in Table 7.3. Except for FT-MFAM, the predictive accuracy of all modeling methods increased with the number of HF samples. It can be observed that for low-dimensional problems, the HDR-DTW method achieved similar high accuracy to assumption-based methods (RBF-MFAM, Kriging-MFAM, and DNN-MFAM) and fine-tuning based FT-MFAM. As the dimensionality increased, the predictive capabilities of RBF-MFAM, Kriging-MFAM, and DNN-MFAM fell below those of FT-MFAM and HDR-DTW. This indicates that assumption-based modeling methods perform poorly in high-dimensional problems with limited HF samples and complex model relationships. It also demonstrates the effectiveness of deep transfer learning in multi-fidelity surrogate modeling for high-dimensional complex models with scarce HF samples.

Table 7.3 Accuracy comparison of different surrogate models

Example	Sample (LF,HF)	Target	RBF-MFAM	Kriging-MFAM	DNN-MFAM	FT-MFAM	HDR-DTW
F1	(100, 10)	NRMSE	0.3398	0.3728	0.2688	**0.2764**	0.3301
		R^2	0.8212	0.7848	0.8881	**0.8817**	0.8312
	(100, 20)	NRMSE	0.2391	0.2418	0.2284	**0.1930**	0.1931
		R^2	0.9115	0.9095	0.9192	**0.9423**	0.9422
F2	(100, 10)	NRMSE	0.1810	0.3869	0.2591	**0.0737**	0.1835
		R^2	0.9492	0.7682	0.8961	**0.9915**	0.9478
	(100, 20)	NRMSE	0.1216	0.1911	0.1857	**0.1133**	0.1796
		R^2	0.9770	0.9434	0.9466	**0.9801**	0.9500
F3	(100, 10)	NRMSE	0.1356	0.1069	0.1302	0.1260	**0.0689**
		R^2	0.9146	0.9282	0.9213	0.9262	**0.9779**
	(100, 20)	NRMSE	0.0678	0.0858	0.0923	0.0955	**0.0558**
		R^2	0.9786	0.9657	0.9604	0.9576	**0.9855**
F4	(150, 15)	NRMSE	0.0865	0.0981	0.0863	0.0710	**0.0698**
		R^2	0.9305	0.9108	0.9308	0.9412	**0.9547**
	(150, 30)	NRMSE	0.0489	0.0570	0.0474	0.0443	**0.0437**
		R^2	0.9777	0.9698	0.9791	0.9817	**0.9822**
F5	(200, 20)	NRMSE	**0.0353**	0.1572	0.0956	0.0400	0.0730
		R^2	**0.9976**	0.9541	0.9830	0.9970	0.9901
	(200, 40)	NRMSE	**0.0345**	0.0804	0.0629	0.0638	0.0664
		R^2	**0.9978**	0.9880	0.9926	0.9924	0.9918
F6	(300, 30)	NRMSE	0.3085	0.2971	0.2921	**0.1661**	0.2095
		R^2	0.8278	0.8403	0.8457	**0.9501**	0.9206
	(300, 60)	NRMSE	0.2550	0.2451	0.2302	0.1604	**0.1149**
		R^2	0.8823	0.8913	0.9041	0.9534	**0.9761**
F7	(300, 30)	NRMSE	0.2796	0.2497	0.4182	**0.1976**	0.2060
		R^2	0.5902	0.6733	0.0835	**0.7954**	0.7775
	(300, 60)	NRMSE	0.2442	0.2387	0.4098	0.2061	**0.1789**
		R^2	0.6873	0.7012	0.1201	0.7772	**0.8322**
F8	(400, 40)	NRMSE	0.2008	0.2129	0.2047	0.1706	**0.1207**
		R^2	0.8384	0.8185	0.8321	0.8834	**0.9415**
	(400, 80)	NRMSE	0.1781	0.1657	0.1323	0.0934	**0.0778**
		R^2	0.8728	0.8899	0.9298	0.9650	**0.9757**
F9	(800, 80)	NRMSE	0.2456	0.2488	0.2436	**0.2014**	0.2296
		R^2	0.4784	0.4651	0.4875	**0.6493**	0.5443
	(800, 160)	NRMSE	0.2326	0.2280	0.2237	0.2005	**0.1991**
		R^2	0.5323	0.5507	0.5675	0.6524	**0.6574**
F10		NRMSE	0.2294	0.2433	0.2283	0.1892	**0.1858**
	(1000, 100)	R^2	0.4246	0.3530	0.4300	0.6087	**0.6225**
	(1000, 200)	NRMSE	0.2202	0.2272	0.2134	0.1922	**0.1841**
		R^2	0.4700	0.4359	0.5023	0.5960	**0.6295**

Comparing the two transfer learning-based methods (FT-MFAM and HDR-DTW), HDR-DTW outperformed FT-MFAM in seven out of ten problems. Notably, for certain problems, increasing the number of HF samples did not enhance the accuracy of FT-MFAM and even led to a decline. This is attributed to the potential overfitting of the last few layers of the neural network during fine-tuning, which impairs the model's generalization ability. Overfitting results in good performance on the training set, but significantly reduced capability to handle new data, ultimately undermining the model's overall performance. In contrast, the proposed HDR-DTW

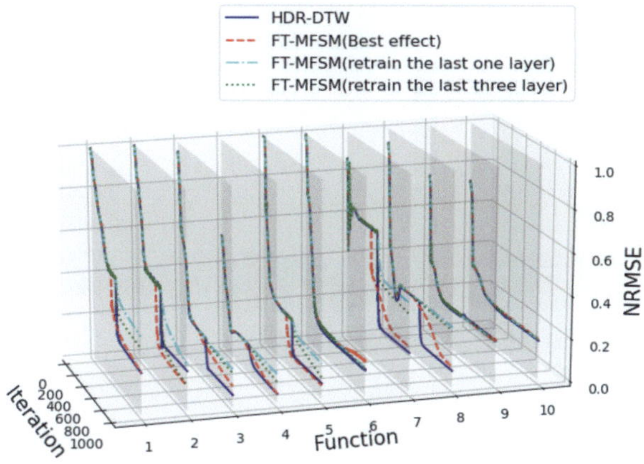

Fig. 7.8 Comparison of NRMSE convergence curves between FT-MFAM and HDR-DTW

employs a dynamic adjustment process that gradually amplifies the influence of HF samples during training. This allows the model to perform well not only on training data, but also to exhibit stronger generalization ability and higher accuracy on new data.

From the comparison of NRMSE convergence curves between FT-MFAM and HDR-DTW in Fig. 7.8, it is evident that the accuracy of FT-MFAM is markedly affected by the choice of fine-tuning layers, leading to performance fluctuations.

To further compare the predictive performance of the five MF surrogate modeling methods, the mean and standard deviation of $NRMSE$ and R^2 were calculated based on the evaluation index results in Table 7.3, reflecting overall prediction accuracy and robustness, respectively. Robustness indicates the model's anti-interference capability, i.e., consistent predictive performance across different test cases or sample sizes. Figure 7.9 presents a comprehensive assessment of the five MF surrogate models. As shown in Fig. 7.9a, the DNN-MFAM method demonstrates the lowest overall prediction accuracy, followed by RBF-MFAM and Kriging-MFAM, which exhibit similar accuracy but are still inferior to FT-MFAM. Among all methods, HDR-DTW achieves the highest overall prediction accuracy. Figure 7.9b reveals that all five methods exhibit small standard deviations, indicating high robustness. Among them, HDR-DTW shows the lowest standard deviation, suggesting the best overall robustness. FT-MFAM, RBF-MFAM, and Kriging-MFAM also exhibit comparable robustness, while DNN-MFAM has the poorest overall robustness, significantly lagging behind the other methods. In summary, the model constructed using HDR-DTW demonstrates the best predictive performance, validated through ten test cases, confirming the effectiveness of the proposed method.

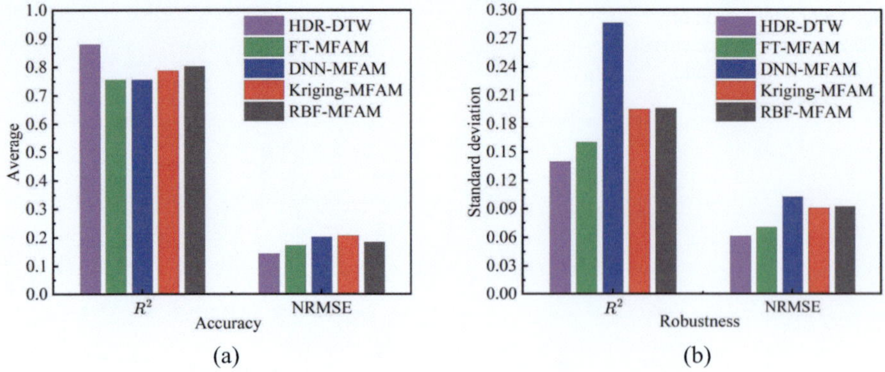

Fig. 7.9 Performance comparison of different models: **a** accuracy, **b** robustness

7.2.3 Discussion on Dynamic Task-Weighting Strategy

To further explore and validate the effectiveness of the dynamic weighting loss strategy, ablation experiments were conducted. Following the examples and hyperparameter settings provided in Table 7.2, experiments compared the performance differences between fixed and dynamic weighting strategies during model training.

In the fixed weighting strategy, several weight combinations were tested for the source domain loss L_1 and target domain loss L_2. Specifically, three weight settings were examined: $\alpha = 1 : \beta = 1$, where the source and target domain losses have equal weights; $\alpha = 1.5 : \beta = 0.5$, where the source domain loss weight is slightly higher than the target domain loss weight; and $\alpha = 0.5 : \beta = 1.5$, where the target domain loss weight exceeds that of the source domain loss.

From the convergence curves in Fig. 7.10, it is evident that networks trained with the dynamic weighting loss strategy exhibit superior performance. In contrast, networks trained with fixed weighting strategies show loss curves that quickly plateau, potentially indicating gradient vanishing issues.

To further validate this observation, the performance of models trained with the dynamic weighting loss strategy was compared against models trained with three fixed weighting strategies on a validation set. The comparison results are presented in Table 7.4. It can be seen that models trained with the dynamic weighting loss strategy achieve significantly higher accuracy than those trained with fixed weighting strategies. This result not only confirms the effectiveness of the dynamic weighting loss strategy but also demonstrates that it can effectively pre-

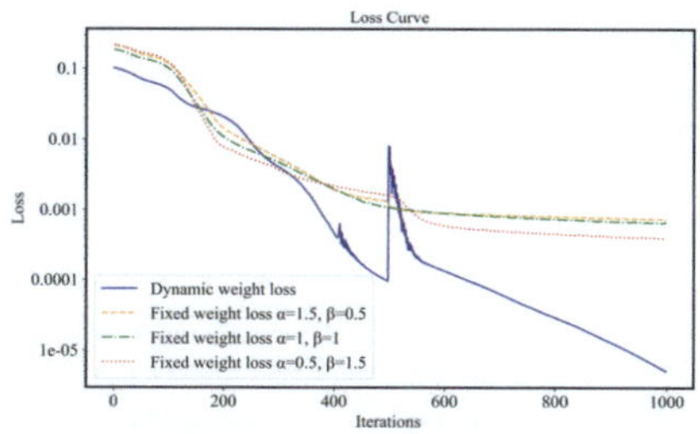

(a) F1 Error convergence curve

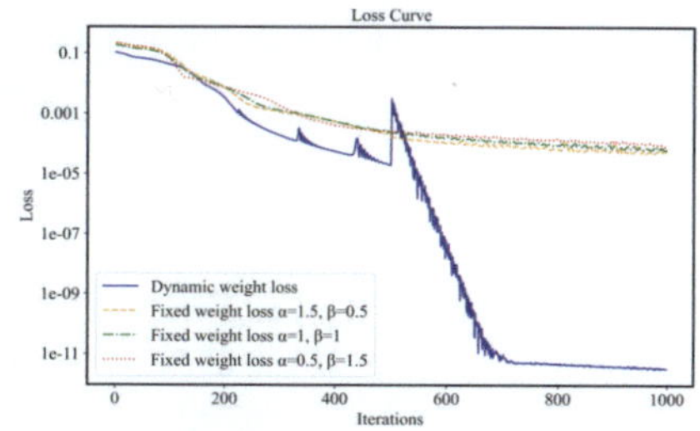

(b) F2 Error convergence curve

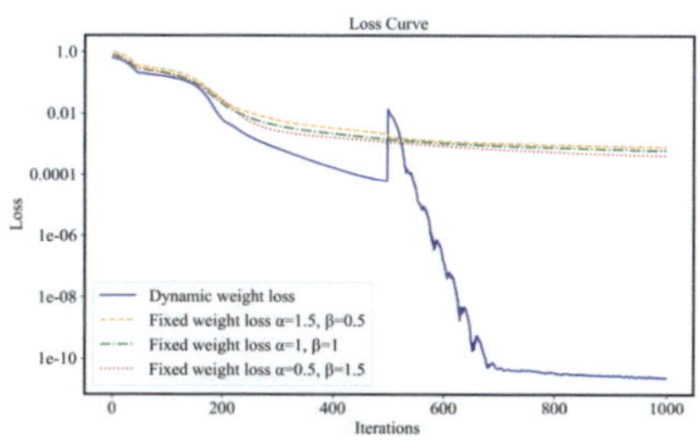

(c) F3 Error convergence curve

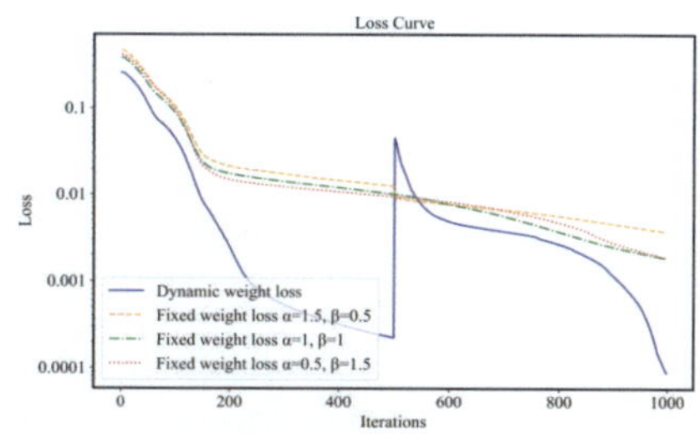

(d) F4 Error convergence curve

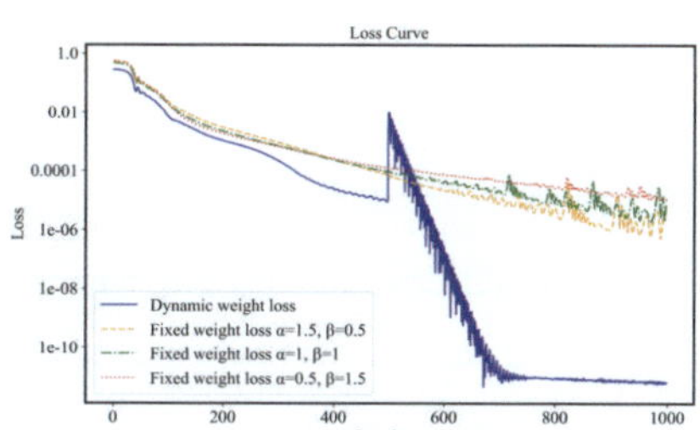

(e) F5 Error convergence curve

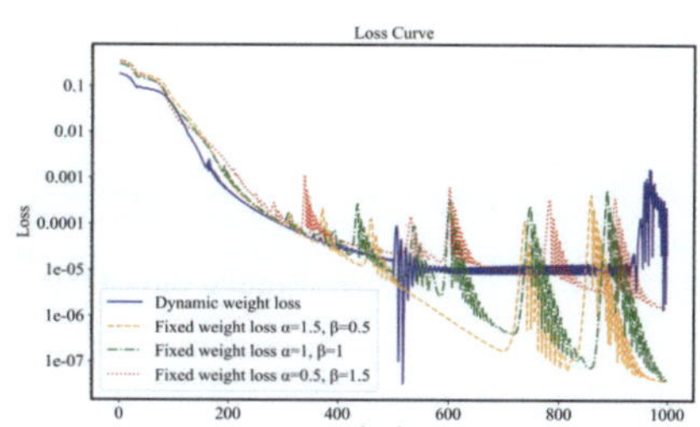

(f) F6 Error convergence curve

Fig. 7.10 Comparison of convergence speeds under different weighting strategies

vent single-task dominance during joint training, reducing the likelihood of network overfitting and enabling more balanced and efficient learning (Figs. 7.11 and 7.12, Tables 7.5 and 7.6).

Table 7.4 Accuracy comparison between dynamic and fixed weighting (NRMSE)

			Fixed weight $L = \alpha L_1 + \beta L_2$		
Example	Dimension	Dynamic weight	$\alpha, \beta = 1, 1$	$\alpha, \beta = 1.5, 0.5$	$\alpha, \beta = 0.5, 1.5$
F1	2	**0.1931**	0.4444	0.4423	0.4661
F2	2	**0.1796**	0.4132	0.4113	0.4036
F3	2	**0.0558**	0.1415	0.1565	0.1275
F4	3	**0.0437**	0.1088	0.1185	0.9525
F5	4	**0.0664**	0.2169	0.2000	0.2171
F6	6	**0.1149**	0.1256	0.1179	0.1357
F7	6	**0.1789**	0.5764	0.5818	0.5680
F8	8	**0.0778**	0.2489	0.2506	0.2492
F9	16	**0.1991**	0.2029	0.2033	0.1997
F10	20	**0.1841**	0.1862	0.1861	0.1866

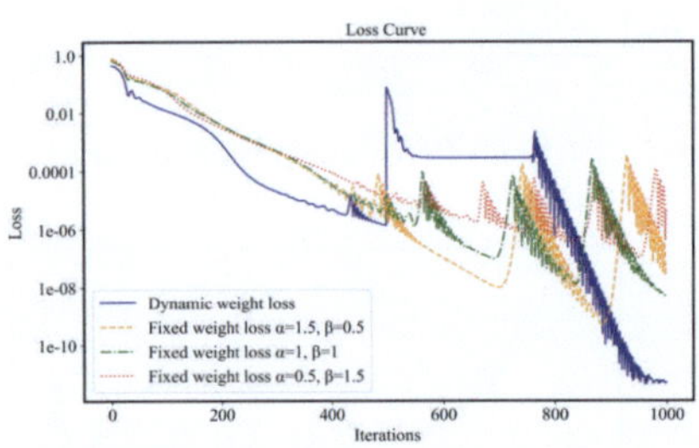

(a) F7 Error convergence curve

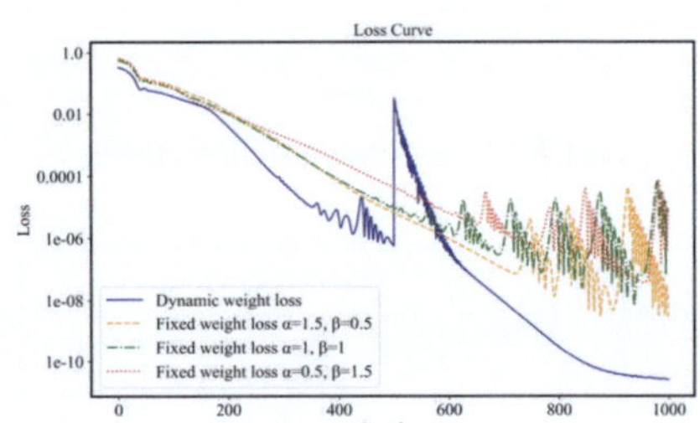

(b) F8 Error convergence curve

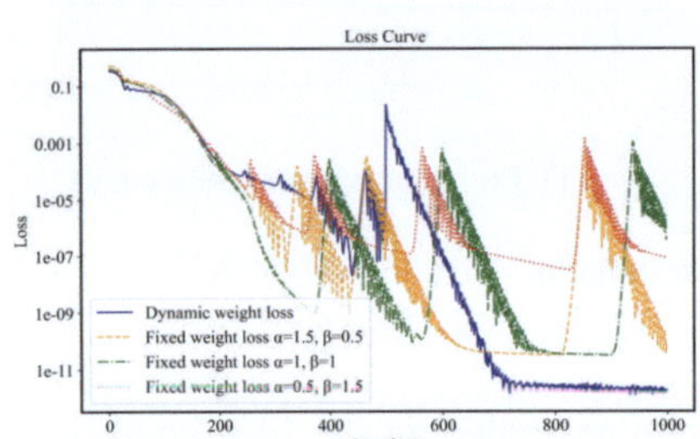

(c) F9 Error convergence curve

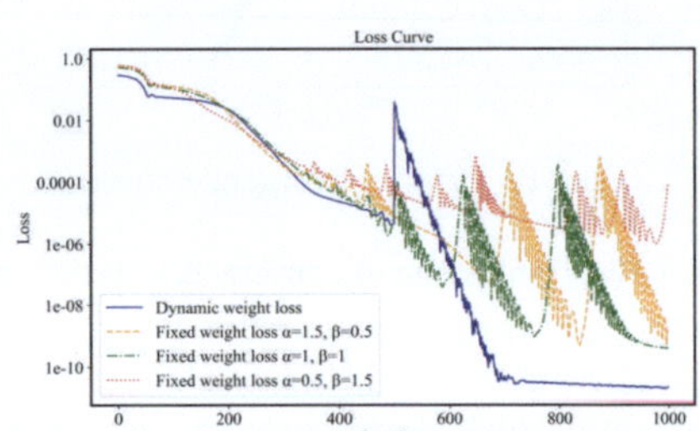

(d) F10 Error convergence curve

Fig. 7.11 Comparison of convergence speeds under different weighting strategies (continued)

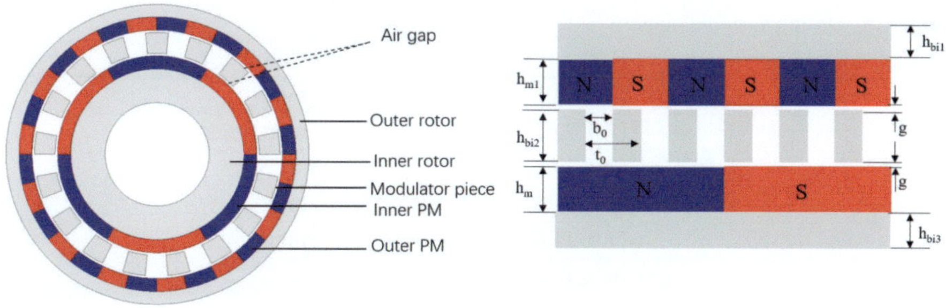

(a) Coaxial PM gear configuration (b) Linearized CMG model

Fig. 7.12 Coaxial permanent magnet gear

Table 7.5 CMG design parameters

Parameter	Symbol	Value	Unit
Number of inner rotor pole pairs	p_h	3	–
Number of outer rotor pole pairs	p_l	14	–
Number of stationary steel pole pieces	N_s	17	–
Iron thickness of stationary steel	h_{bi2}	8.0	mm
External diameter	D_e	120	mm
Shaft diameter	D_{sh}	40	mm
Stack length	L_{stk}	100	mm
Air gap length	g	1.0	mm
Yoke thickness of inner rotor	h_{bi3}	13.0	mm
Yoke thickness of outer rotor	h_{bi1}	5.0	mm
PM thickness of inner rotor	h_m	7.0	mm
PM thickness of outer rotor	h_{m1}	5.0	mm
Inner magnet-arc to PM pole pitch ratio	α	1.0	–
Outer magnet-arc to PM pole pitch ratio	α_1	1.0	–
Air slot opening to pole pitch ratio	c_0	0.5	–
Residual flux density of PMs	B_r	1.25	T
Relative differential permeability of PMs	μ_r	1.077	–

7.3 Engineering Application: Optimization Design of Permanent Magnet Gear

$$\textbf{Max:} \quad f(h_{bi1}, \alpha_1, h_{m1}, h_{bi3}, h_m, \alpha, h_{bi2}, c_0) = Torque. \tag{7.14}$$

7.3.1 Constructing MF Surrogate Models with HDR-DTW

In this study, ANSYS Maxwell 2021 R1 commercial software was employed for torque analysis of the coaxial magnetic gear (CMG). Two finite-element models

Table 7.6 Design variable ranges

Design variable	Range	Unit
h_{bi1}	[1.0, 7.0]	mm
h_{m1}	[1.0, 7.0]	mm
α_1	[0.7, 1.0]	–
h_{bi3}	[5.5, 14.5]	mm
h_m	[7.0, 11.0]	mm
α	[0.7, 1.0]	–
h_{bi2}	[4.0, 11.0]	mm
c_0	[0.5, 0.52]	–

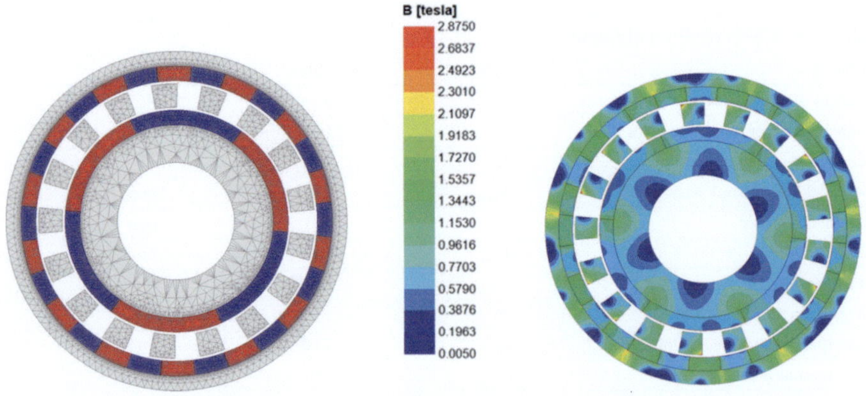

(a) Grid model. (b) Simulation analysis.

Fig. 7.13 2D Maxwell model with 4 mm mesh (LF model)

were utilized: a 2D LF model with a 4mm mesh (see Fig. 7.13) and a 3D HF model with a 15 mm mesh (see Fig. 7.14). The HF and LF simulations for the CMG required approximately 900 and 60 s, respectively, indicating that 15 LF simulations equate to the time of one HF simulation. To evaluate the effectiveness of HDR-DTW, 250 LF and 30 HF samples were generated using the LHS method, and the corresponding torques were obtained via ANSYS Maxwell software.

HDR-DTW was trained using the generated samples, with a learning rate set to 1×10^{-4}. Bayesian optimization (BO) was employed for hyperparameter tuning, with a maximum of 100 iterations. The optimized hyperparameters are presented in Table 7.7. Figure 7.15 illustrates the predictive accuracy of HDR-DTW compared to Co-RBF, Co-Kriging, DGP-MFSM, DNN-MFSM, and FTL-MFSM when using the same number of HF and LF samples. It can be observed that the NRMSE of HDR-DTW is 65.04, 56.55, 68.34, 73.40, and 53.04% lower than those of Co-RBF, Co-Kriging, DGP-MFSM, DNN-MFSM and FTL-MFSM, respectively.

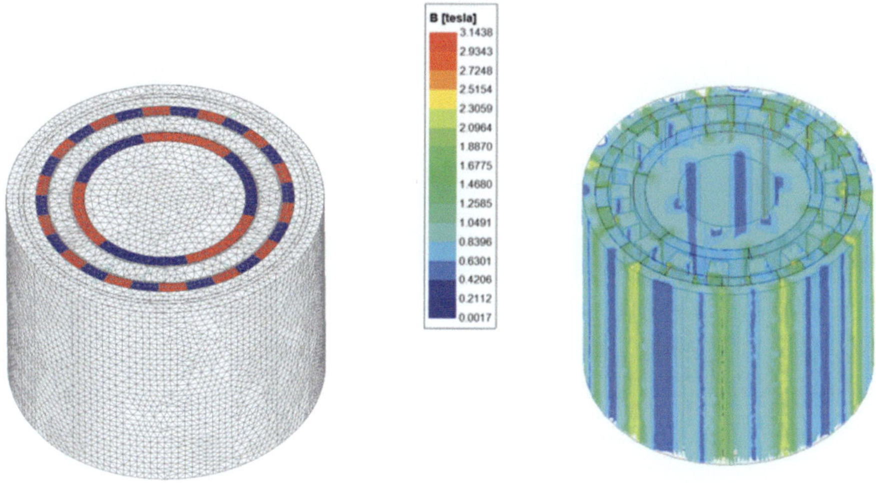

(a) Grid model (b) Simulation analysis

Fig. 7.14 3D Maxwell model with 15 mm mesh (HF model)

Table 7.7 Optimized hyperparameters of HDR-DTW (30 HF + 250 LF) for the CMG case

Hyperparameter	Lower bound	Upper bound	Optimized result with BO
Number of neurons in encoder layer	20	300	232
Number of hidden layers in encoder	2	9	4
Number of neurons in classifier layer	20	300	75
Number of hidden layers in classifier	2	9	8
Gradient threshold	0.2	5	0.037778

Compared to single-fidelity surrogate modeling methods, the computational efficiency of HDR-DTW is also of interest. Therefore, the prediction accuracy of single-fidelity surrogates based solely on HF samples was investigated by incrementally adding HF samples from 60 to 200. The prediction accuracies are listed in Table 7.8. As the number of HF samples increases, there is a clear upward trend in the prediction accuracy of single-fidelity surrogate models. To achieve the prediction accuracy of HDR-DTW (30 HF + 250 LF), HF-DNN requires 190 HF samples, Kriging needs 80 HF samples, and RBF requires 170 HF samples.

As mentioned earlier, the time spent on 15 LF samples is equivalent to that of one HF sample. Consequently, compared to single-fidelity surrogate models such as HF-DNN, Kriging, and RBF, HDR-DTW reduces the total computational time by 75.44, 41.67, and 72.55%, respectively. In other words, it significantly saves computational costs, aligning with the primary concept of multi-fidelity surrogate

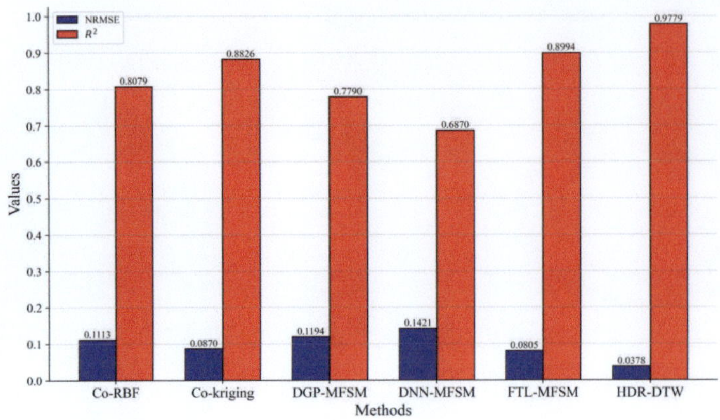

Fig. 7.15 Predictive Accuracy of MFSMs (30 HF + 250 LF) for CMG

Table 7.8 Predictive accuracy of CMG

		60 HF	100 HF	120 HF	150 HF	200 HF	Required HF samples
	Computing cost (h)	15	25	30	37.5	50	
HF-DNN	NRMSE R^2	0.0889 0.9254	0.0821 0.8992	0.0434 0.9610	0.0438 0.9727	0.0345 0.9766	190
Kriging	NRMSE R^2	0.0438 0.9818	0.0279 0.9882	0.0229 0.9890	0.0202 0.9941	0.0082 0.9986	80
RBF	NRMSE R^2	0.0981 0.9091	0.1099 0.8193	0.0615 0.9218	0.0584 0.9514	0.0269 0.9857	180

modeling. Thus, the engineering case of the CMG demonstrates that the proposed HDR-DTW achieves high prediction accuracy and efficiency.

7.3.2 Optimization Design Based on Genetic Algorithm

Based on the aforementioned HDR-DTW, a single-objective genetic algorithm (GA) was employed to search for optimal parameters to maximize the torque of the CMG. The output of HDR-DTW served as the individual fitness. The GA parameters were configured as follows: population size of 100, maximum generations of 500, crossover rate of 0.8, and mutation rate of 0.2. The initial and optimal design parameters of the CMG are presented in Table 7.9. Finite element simulation validation was also conducted. It can be observed that after GA optimization using HDR-DTW, the output torque increased from 115 N.m in the initial design to 168.91 N.m, representing a 46.87% improvement. Furthermore, the entire process consumed only 30 HF

Table 7.9 Optimal results of CMG

Design variable	Initial parameter	Optimal parameter	FEA validation	Unit
h_{bi1}	5.0	6.55	6.55	mm
h_{bi2}	8	11	11	mm
h_{bi3}	13	14.5	14.5	mm
h_{m1}	5.0	7.0	7.0	mm
h_m	7	11	11	mm
α_1	1.0	1.0	1.0	–
α	1.0	1.0	1.0	–
c_0	0.5	0.5169	0.5169	–
Torque	115	168.91	174.52	N.m

and 250 LF samples, significantly reducing computational costs during design optimization. Table 7.9 also indicates that the torque values from HDR-DTW evaluation and high-fidelity finite-element evaluation differ only by 3.21%, which is acceptable.

7.4 Summary

This chapter presented HDR-DTW, a multi-fidelity transfer learning model that synergizes harmonic domain regression with dynamic task-weighted loss. The framework's core innovation lies in its dual-phase training strategy: an initial feature extraction phase prioritizes LF data to distill cross-fidelity representations, while a subsequent regression phase shifts focus to HF accuracy through adaptive loss weighting. This dynamically balanced approach mitigates gradient conflicts, convergence imbalances, and task dominance-critical limitations in conventional joint-training methods.

Comprehensive validation across ten benchmark functions revealed HDR-DTW's superiority over four established MFAM techniques (RBF-MFAM, Kriging-MFAM, DNN-MFAM, and FT-MFAM). Notably, it achieved 12–68% lower NRMSE and 15–40% higher R^2 in high-dimensional problems, demonstrating exceptional robustness with standard deviations below 0.05. The dynamic weighting strategy proved essential, outperforming fixed-weight alternatives by 22–53% in prediction accuracy.

In the coaxial magnetic gear design case, HDR-DTW reduced computational costs by 41–75% compared to single-fidelity surrogates while maintaining torque prediction errors below 3.2%. The optimized gear design enhanced torque output by 46.87%, validated through finite-element analysis. These results confirm

HDR-DTW's efficacy in overcoming scarce HF data limitations and complex fidelity relationships, establishing it as a robust framework for data-efficient engineering optimization.

References

1. McKay MD, Beckman RJ, Conover WJ (2000) A comparison of three methods for selecting values of input variables in the analysis of output from a computer code. In: Technometrics 42(1):55–61 (2000)
2. Yi J, Li X, Xiao M et al (2017) Construction of nested maximin designs based on successive local enumeration and modified novel global harmony search algorithm. Eng Optim 49(1):161–180
3. Meng X, Karniadakis GE (2020) A composite neural network that learns from multi-fidelity data: application to function approximation and inverse PDE problems. J Comput Phys 401:109020

Chapter 8
Online Variable-Fidelity Surrogate-Assisted Harmony Search Algorithm with Multi-level Screening Strategy

In the realm of computationally expensive design optimization, where high-fidelity simulations often demand prohibitive resources, surrogate-assisted methods have emerged as indispensable tools for balancing accuracy and efficiency. This chapter advances this paradigm by integrating variable-fidelity surrogate models that fuse data from multiple fidelity levels to approximate complex objective functions, with a metaheuristic optimizer, specifically the Harmony Search (HS) algorithm. Originally proposed by Geem in 2001 as a music-inspired metaheuristic [1], HS excels in global exploration through iterative improvisation of solution "harmonies", making it suitable for diverse applications such as robotic path planning [2, 3], multi-axis machining [4], and renewable energy system design [5, 6]. Yet, its heavy reliance on numerous function evaluations limits its utility in high-cost scenarios, underscoring the need for surrogate augmentation.

To bridge this gap, we present the Variable-Fidelity Surrogate-Assisted Harmony Search (VFS-HS) algorithm, a novel framework that embeds adaptive variable-fidelity surrogates within the HS process to drastically reduce high-fidelity evaluations while preserving optimization fidelity. By constructing surrogates that leverage abundant low-fidelity data to inform sparse high-fidelity samples, VFS-HS enhances predictive accuracy and guides the search toward promising regions. A pivotal contribution is the multi-level screening strategy, rooted in non-dominated sorting, which dynamically filters candidates across fidelity tiers to optimize computational allocation and harmonize exploration with exploitation. Comprehensive assessments on the CEC2014 expensive optimization benchmarks, benchmarked against leading surrogate-assisted techniques (OPUS-RBF, TLSAPSO, SHPSO, HSOSR, MGP-MDE), affirm VFS-HS's superior performance in convergence speed and solution quality. This is further validated through a practical engineering application: the structural optimization of a long cylindrical pressure vessel, illustrating VFS-HS's efficacy in real-world, high-stakes design problems. Through this lens, the chapter elucidates how variable-fidelity surrogates elevate metaheuristic optimization, aligning with the monograph's broader emphasis on experimental design, modeling, and applications in surrogate-based frameworks.

J. Yi et al., *Variable-Fidelity Surrogate*, Engineering Applications of Computational Methods 25, https://doi.org/10.1007/978-981-95-5527-7_8

8.1 Variable-Fidelity Surrogates-Assisted Harmony Search Algorithm

8.1.1 Improved Harmony Search Algorithm (MHS)

The classical Harmony Search (HS) algorithm is inefficient in performing local searches in numerical optimization applications and is sensitive to parameter settings (HMS, $HMCR$, and PAR) [7]. To enhance its performance, we proposed an improved HS (MHS) in our previous research, which has demonstrated good performance in function optimization and engineering design optimization problems [8, 9]. Therefore, this study adopts MHS as the basic optimizer. MHS introduces a new pitch adjustment rule with crossover and mutation operations. First, all harmony vectors are divided into two groups based on fitness evaluation: the better group and the worse group. The proportion of the better group is defined as M, where $0 < M < 1$. Next, two different mutation operations are applied to the harmony vectors in the better and worse groups, respectively. For the better group, one harmony (wr_1) is selected from the worse group, and two harmonies (br_1 and br_2) are selected from the better group for mutation, as shown in the following equation:

$$x_i = x_{wr_1} + F \cdot (x_{br_1} - x_{br_2}), \quad \text{where } br_1 \neq br_2 \neq wr_1 \neq i. \tag{8.1}$$

In this equation, F (where $0 < F < 1$) is the mutation parameter. For the worse group, one harmony (br_1) is selected from the better group, and two harmonies (wr_1 and wr_2) are selected from the worse group for mutation, as shown in the following equation:

$$x_i = x_{br_1} + F \cdot (x_{wr_1} - x_{wr_2}), \quad \text{where } wr_1 \neq wr_2 \neq br_1 \neq i. \tag{8.2}$$

To facilitate the implementation and understanding of MHS, Algorithm 8.1 provides detailed pseudocode that explains the steps for improvising new harmonies.

8.1.2 Evaluation Point Selection Criterion Based on Non-dominated Sorting

In the field of surrogate-based optimization, the selection strategy for evaluation points has a decisive impact on algorithm performance. An inappropriate selection strategy may cause the algorithm to fall into local optima or result in inefficient global search capabilities. This paper delves into the core concepts of evaluation point selection strategies and, by introducing theoretical backgrounds and practical cases from related fields, proposes a selection strategy based on non-dominated sorting to achieve a balance between exploration and exploitation.

Algorithm 8.1: Pseudo code of improvising new harmonies in MHS algorithm.

```
Begin
for i = 1 : HMS do
    for j = 1 : d do
        if r_1 < HMCR then
            x_i^j = HM(a, j), a ∈ {1, 2, ..., HMS} /* Randomly choose a value
                from HM                                                    */
            if r_2 < PAR then
                if i < M * HMS then
                    x_i^j = x_{wr_1}^j + F · (x_{br_1}^j − x_{br_2}^j), wr_1 ≠ br_1 ≠ br_2 ≠ i /* Intersect
                        mutation operation 1                               */
                else
                    x_i^j = x_{br_1}^j + F · (x_{wr_1}^j − x_{wr_2}^j), br_1 ≠ wr_1 ≠ wr_2 ≠ i /* Intersect
                        mutation operation 2                               */
        else
            x_i^j = LB^j + r_3 · (UB^j − LB^j) /* Randomly generate a variable
                */
```

The choice of sample size is a critical issue in surrogate model optimization. An overly small sample size may lead to insufficient prediction accuracy of the surrogate model, thereby affecting the accuracy of the optimization process. Conversely, an excessively large sample size may result in the wastage of computational resources. Sequential sampling strategies offer a solution, with the core idea being to incrementally add new sample points to enhance the precision of the surrogate model. The crux of this strategy lies in the selection of evaluation points, which directly impacts the algorithm's performance.

Evaluation point selection strategies can be divided into two categories: exploitation-biased and exploration-biased. Exploitation-biased strategies tend to add new sample points near the current optimal value of the surrogate model to improve local prediction accuracy. However, this approach may lead to an overconcentration of sampling points, causing the algorithm to fall into local optima and waste computational resources. In contrast, exploration-biased strategies tend to add new sample points in sparsely sampled regions to enhance the global prediction accuracy of the surrogate model. However, this strategy may result in insufficient local accuracy due to a lack of adequate sampling in regions with high nonlinearity of the true function.

To achieve a balance between exploration and exploitation, this paper proposes an evaluation point selection strategy based on non-dominated sorting. Non-dominated sorting is a core concept in multi-objective optimization. For a multi-objective optimization problem with m conflicting objectives, given two solutions a and b ($a, b \in \mathbb{R}^n$), we say that a dominates b if and only if $f_i(a) \leq f_i(b)$ for all $i \in [1, 2, \ldots, m]$, and there exists at least one k ($k \in [1, 2, \ldots, m]$) such that

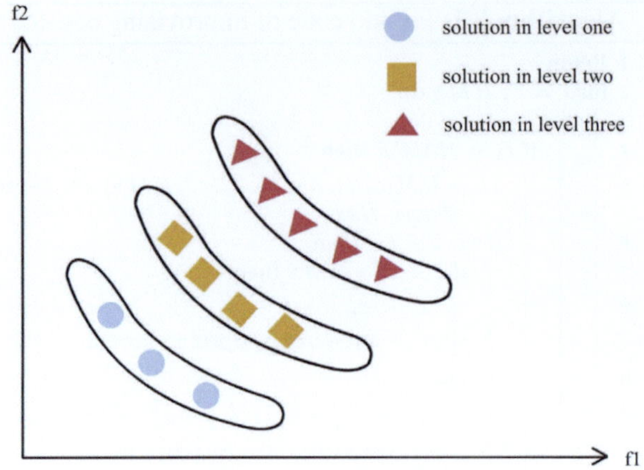

Fig. 8.1 Illustration of different non-dominated levels

$f_k(a) < f_k(b)$. Non-dominated sorting involves ranking solutions based on their mutual relationships into different non-dominated levels, as shown in Fig. 8.1. It can be seen that solutions in the first non-dominated level dominate solutions in other levels.

The main idea of the evaluation point selection strategy based on non-dominated sorting is to treat exploration and exploitation as two mutually constraining objectives and then use non-dominated sorting to select potential evaluation points. The primary challenge of this method is to objectively evaluate the capabilities of exploration and exploitation. Exploitation involves performing local search around the current position x, and thus can be assessed using the estimated objective function value from the variable-fidelity surrogate model. Exploration capability, on the other hand, refers to the ability to explore unknown regions and can be evaluated by the sum of distances between the position and other positions in the population [10].

8.1.3 VFS-HS Model Framework

The flowchart of VFS-HS is shown in Fig. 8.2. The detailed steps are as follows:

Step 1 Nested Experimental Design. Generate a nested sampling set $D = \{D_l, D_h\}$, where $D_l = \mathbf{x}_1, \mathbf{x}_2, \ldots, \mathbf{x}_{n_L}$ and $D_h = \mathbf{x}_1, \mathbf{x}_2, \ldots, \mathbf{x}_{n_H}$. D_h is nested within D_l ($D_h \subset D_l$), with D_l containing a large number of sampling points for evaluating the low-fidelity (LF) model, and D_h containing a small number of sampling points for evaluating the high-fidelity (HF) model. Here, we use the method proposed by Yi et al. [11] to generate the nested sampling set.

Step 2 Run the Low-Fidelity Model. Evaluate the low-fidelity responses $f_l(\mathbf{x}_i)$ for $\mathbf{x}_i \in D_l, i = 1, 2, \ldots, n_L$. Set the number of low-fidelity model evaluations $LFEs$ to n_L. Initialize the database Θ_l and set $\Theta_l = \{(\mathbf{x}_i, f_l(\mathbf{x}_i)), i = 1, 2, \ldots, LFEs\}$.

Step 3 Run the High-Fidelity Model. Evaluate the high-fidelity responses $f_h(\boldsymbol{x}_k)$ for $\boldsymbol{x}_k \in D_h,\ k = 1, 2, \ldots, n_H$. Set the number of high-fidelity model evaluations $HFEs$ to n_H. Then initialize the database Θ_h and set $\Theta_h = \{(\boldsymbol{x}_k, f_h(\boldsymbol{x}_k)), k = 1, 2, \ldots, HFEs\}$.

Step 4 Construct the Low-Fidelity Surrogate. Use the database Θ_l to build the surrogate $S_L(x)$ for the low-fidelity model.

Step 5 Construct the Variable-Fidelity Surrogate Model. Compare the differences between the low-fidelity and high-fidelity models at the overlapping sampling points $\boldsymbol{x}_k \in D_h, k = 1, 2, \ldots, n_H$. Then construct the variable-fidelity surrogate model $S_V(\boldsymbol{x})$ according to the steps shown in Fig. 8.2.

Step 6 Initialize the Harmony Memory. Use D_h to form the initial harmony memory $HM = \{(\boldsymbol{x}, f_h(\boldsymbol{x})) : \boldsymbol{x} \in D_h\} = \{(\boldsymbol{x}_k, f_h(\boldsymbol{x}_k)) : k = 1, 2, \ldots, n_H\}$, and set $HMS = n_H$. Let $\boldsymbol{x}_b$ be the current best harmony, with $f_b = f_h(\boldsymbol{x}_b)$.

Step 7 Improvise New Harmonies. For each harmony in the current harmony memory, generate a candidate solution according to the improvisation process shown in Fig. 8.2, and evaluate all candidate solutions using the variable-fidelity surrogate model (without low-fidelity evaluations) $\hat{S}_V(\boldsymbol{x})$. $\hat{S}_V(\boldsymbol{x})$ uses $S_L(\boldsymbol{x})$ instead of the actual low-fidelity model to represent the low-fidelity part.

Step 8 Non-dominated Sorting. Obtain the first non-dominated level of candidates $D_n = \mathbf{x}_1, \mathbf{x}_2, \ldots, \mathbf{x}_m$ using the non-dominated sorting method. Re-evaluate the selected candidates D_n using the variable-fidelity surrogate $S_V(\mathbf{x})$ and low-fidelity evaluations. Update the number of low-fidelity evaluations $LFEs$ to: $LFEs = LFEs + m$. Use non-dominated sorting again to select $s (s \leq m)$ candidates as the first non-dominated level, and re-evaluate these s candidates using the actual high-fidelity model. Update the harmony memory and the current best harmony $\mathbf{x}_b$. Update the number of high-fidelity evaluations $HFEs$ to: $HFEs = HFEs + s$.

Step 9 Update Databases and Surrogate Models. Add all previously evaluated low-fidelity (LF) and high-fidelity (HF) points to the databases Θ_l and Θ_h, respectively. Update the variable-fidelity surrogate model $S_V(\boldsymbol{x})$ using all previously overlapping sampling points in Θ_h and Θ_l, and update the low-fidelity surrogate model $S_L(\boldsymbol{x})$ using all previously sampled points in Θ_l.

Step 10 Local Search. Perform a local search around the current best harmony $\mathbf{x}_b$. Use a local optimization solver (e.g., MATLAB's `fmincon`) to obtain the optimal solution $\mathbf{x}_b^*$ within the domain $[\mathbf{x}_b - \mathbf{R}, \mathbf{x}_b + \mathbf{R}] \cap [\mathbf{LB}, \mathbf{UB}]$ using the variable-fidelity surrogate $\hat{S}_V(\mathbf{x})$ (without using LF evaluations). Re-evaluate $\mathbf{x}_b^*$ using the variable-fidelity surrogate $S_V(\mathbf{x})$ and low-fidelity evaluations. Update the number of low-fidelity evaluations $LFEs$ to: $LFEs =$

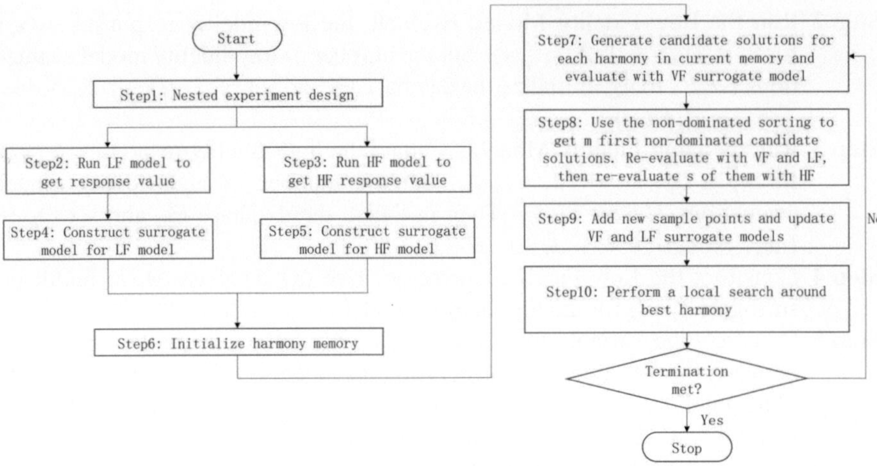

Fig. 8.2 Flowchart of the VFS-HS algorithm

$LFEs + 1$, and add it to the database Θ_l. (The number of low-fidelity evaluations $LFEs$ increases because the evaluation of $S_V(\mathbf{x})$ involves the low-fidelity model.) If it is better than $\mathbf{x}_b$, re-evaluate it using the high-fidelity model. If it still outperforms $\mathbf{x}_b$, update the current best harmony $\mathbf{x}_b$, update the number of high-fidelity evaluations $HFEs$ to: $HFEs = HFEs + 1$, and add it to the database Θ_h.

Step 11 Check Termination Criteria. Several different stopping criteria can be used (e.g., reaching the maximum number of "equivalent" HF model evaluations, or when the current best solution stagnates for several consecutive iterations). If the termination criteria are met, stop the algorithm and output the final solution. Otherwise, return to Step 6.

8.1.4 VFS-HS Model Management Strategy

In the proposed VFS-HS method, three different types of models are used to evaluate candidate solutions: the variable-fidelity surrogate model without LF evaluation ($\hat{S}_V(x)$), the variable-fidelity surrogate model with LF evaluation ($S_V(x)$), and the HF model. In this section, the management of these three types of models in the optimization process is discussed and compared with previous methods.

Table 8.1 presents the properties of the three models in VFS-HS. The HF model has the highest fidelity but is also the most computationally expensive. In contrast, ($\hat{S}_V(x)$) has the lowest fidelity and the lowest computational cost. ($S_V(x)$) lies in the middle in terms of both fidelity and computational cost. The model management

Table 8.1 Comparison of properties of different models in VFS-HS

Model	Fidelity	Computational cost
$(\hat{S}_V(x))$	Low	Low
$(S_V(x))$	Medium	Medium
HF model	High	High

in VFS-HS primarily addresses two issues: (1) how to allocate evaluations among these three models with different fidelities; and (2) when to conduct evaluations.

The first issue is relatively straightforward to address, as most variable-fidelity surrogate-assisted optimization methods adopt a pyramid-shaped mechanism, which involves using a large number of inexpensive evaluations and a small number of expensive evaluations. However, the second issue has several different solutions. Figure 8.3 compares the model management strategies of three different variable-fidelity surrogate-assisted optimization methods. Figure 8.3a illustrates the model management strategy of Particle Swarm Optimization with Adaptive Model Switching (PSO-AMS) [12]. PSO-AMS starts the search with the lowest-fidelity model and switches to the next higher-fidelity model when a predefined adaptive model switching criterion is met. However, one disadvantage of this method is that if there is a significant difference between the LF and HF models (e.g., the global optima of the HF and LF models are located in different regions of the domain), the search cost incurred by the low-fidelity model may be wasted. Figure 8.3b shows the model management strategy of Multi-fidelity Gaussian Process and Radial Basis Function-assisted Memetic Differential Evolution (MGPMDE). It consists of three steps: optimization based on the LF model, data mining, and optimization based on the HF model. In the first step, both the LF model and its surrogate are used, while in the second and third steps, both the HF model and its surrogate are employed. Although real LF evaluations can correct the predictions of the LF surrogate, the low fidelity of the LF model may still mislead the search process. Figure 8.3c describes the model management strategy proposed in this paper for VFS-HS. VFS-HS employs a multi-level screening mechanism to select potential candidate solutions. In each iteration of VFS-HS, a large number of candidates are first evaluated using the inexpensive variable-fidelity surrogate $(\hat{S}_V(x))$, then a moderate number of candidates are selected via non-dominated sorting and evaluated using $(S_V(x))$, and finally, a small number of candidates are selected again via non-dominated sorting and evaluated using the HF model. The proposed VFS-HS management strategy has two main advantages: First, the multi-level screening mechanism strictly controls the number of real LF and HF evaluations and ensures that computational costs are allocated wisely; second, the separation of HF evaluations in each iteration allows for monitoring the search direction throughout the optimization process.

To better understand the evaluation point selection process in VFS-HS, a numerical example from the study by Park et al. [13] is used to illustrate the process step-by-step. In this example, the HF and LF models are defined as

Fig. 8.3 Management strategies of three different multi-fidelity surrogate-assisted optimization methods

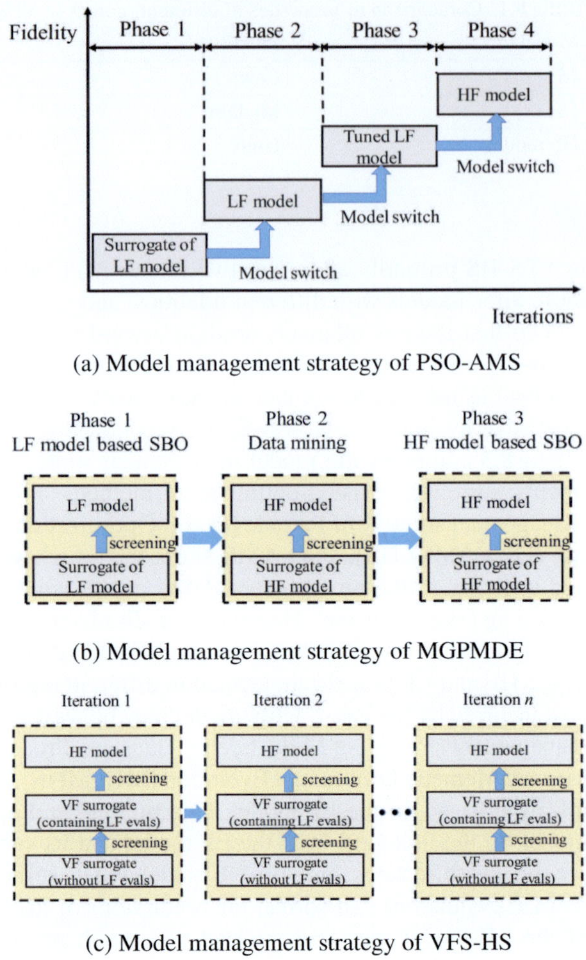

$$y_h(x) = (6x - 2)^2 \sin(12x - 4), \tag{8.3}$$

$$y_l(x) = 0.5y_h(x) + 10(x - 0.5) + 5, \tag{8.4}$$

where $y_h(x)$ represents the HF model, and $y_l(x)$ represents the LF model. Figure 8.4a illustrates these two models. First, a set of nested initial samples (five initial LF samples and three initial HF samples) is generated, and the initial LF and VF surrogate models are established, as shown in Fig. 8.4b. Next, five potential samples (P1–P5) are generated through MHS operations and evaluated using the VF surrogate model without LF evaluation ($\hat{S}_V(x)$), as depicted in Fig. 8.4c. The next step is to select the best points from the five potential samples for real LF evaluation using non-dominated sorting, as shown in Fig. 8.4d. In this step, exploration and exploitation are treated

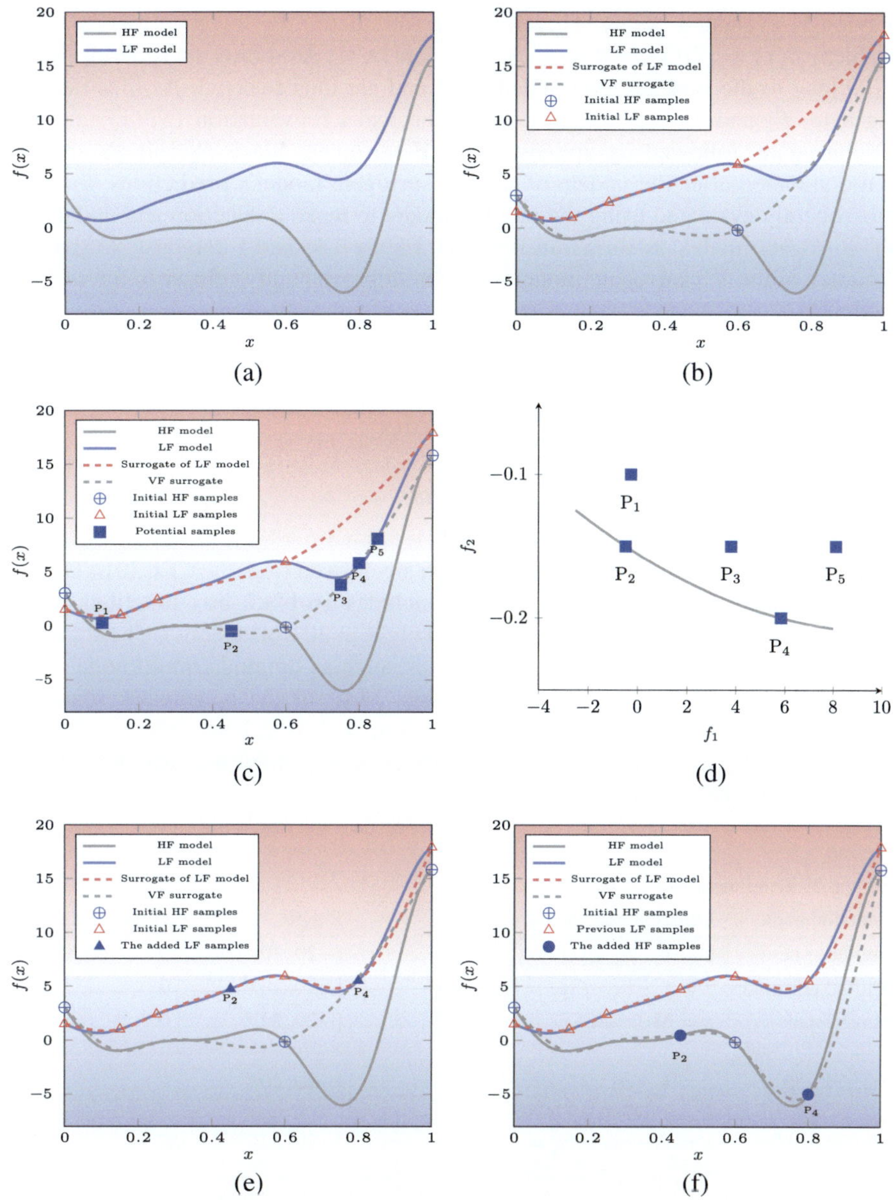

Fig. 8.4 Schematic diagram of the evaluation point selection process

as two conflicting objectives. The exploitation objective ($f1$) is evaluated using the estimated objective function value from ($S_V(x)$). However, the exploration objective ($f2$) can be assessed by the minimum distance to previously evaluated points ($mDis$, using Euclidean distance here), as we aim to select points that are farther away from previously evaluated points. Therefore, we formulate a bi-objective minimization problem $f(x) = \{f_1(x) = S_V(x), f_2(x) = -mDis\}$. As shown in Fig. 8.4d, P_2 and P_4 belong to the first non-dominated level and are thus selected. As illustrated in Fig. 8.4e, P_2 and P_4 are evaluated using the real LF evaluation ($S_V(x)$) and the VF surrogate, followed by updating the LF surrogate. Since P_2 and P_4 are still non-dominated after the update of the VF surrogate model's predictions, they are subsequently evaluated using the real HF model to make a selection and update the VF surrogate model, as shown in Fig. 8.4f. Figure 8.4e and f demonstrate that the accuracy of the VF surrogate model has been improved and is closer to the real HF model.

8.2 Numerical Experimental Results

8.2.1 Introduction to Test Problems

The benchmark functions used in this section are derived from the CEC2014 Expensive Optimization Test Suite [14], which includes 16 black-box functions (eight well-known problems in 10 dimensions and eight in 30 dimensions). These benchmark functions exhibit diverse characteristics, such as unimodal/multimodal, continuous/discrete, and separable/non-separable. Most of the benchmark functions have been subjected to translation and rotation, with translation vectors denoted as $\mathbf{o}_i = [0_{i1}, 0_{i2}, \ldots, 0_{id}]$ and rotation matrices as $\boldsymbol{M}_i$. Table 8.2 lists the detailed information of the benchmark functions.

Table 8.2 Test functions

Function name	Function expression	Range	Global opt.
Shifted sphere	$z = x - o_1,\ f(z) = \sum_{i=1}^{d} z_i^2$	$[-20, 20]^d$	0
Shifted ellipsoid	$z = x - o_2,\ f(z) = \sum_{i=1}^{d} i z_i^2$	$[-20, 20]^d$	0
Shifted and rotated ellipsoid	$z = \mathbf{M}_i(x - o_3),\ f(z) = \sum_{i=1}^{d} x_i^2$	$[-20, 20]^d$	0
Shifted step	$z = x - o_4,\ f(z) = \sum_{i=1}^{d} (\lfloor x_i + 0.5 \rfloor)^2$	$[-20, 20]^d$	0
Shifted Ackley	$z = x - o_5,\ f(z) = 20 + e - 20\exp(-0.2\sqrt{\frac{1}{n}\sum_{i=1}^{n} z_i^2}) - \exp(\frac{1}{n}\sum_{i=1}^{n}\cos(2\pi z_i))$	$[-32, 32]^n$	0
Shifted Griewank	$z = x - o_6,\ f(z) = \frac{1}{4000}\sum_{i=1}^{d} z_i^2 - \prod_{i=1}^{d}\cos(z_i/\sqrt{i}) + 1$	$[-600, 600]^d$	0
Shifted and rotated Rosenbrock	$z = \mathbf{M}_2(2.048(x - o_7)/100),\ f(z) = \sum_{i=1}^{d-1}[100(z_i^2 - z_{i+1})^2 + (z_i - 1)^2]$	$[-20, 20]^d$	0
Shifted and rotated Rastrigin	$z = \mathbf{M}_3(5.12(x - o_8)/20),\ f(z) = \sum_{i=1}^{d}[z_i^2 - 10\cos(2\pi z_i) + 10]$	$[-20, 20]^d$	0

It is important to note that the objective functions of these problems are not truly computationally expensive. However, if these problems share similar landscapes with expensive problems, it is reasonable to expect that the performance of the algorithm on these problems will be comparable to that on expensive problems. Additionally, conducting comparisons on test problems with lower computational costs can save a significant amount of time, allowing for multiple trials and the acquisition of statistical results. In this study, the true functions are regarded as the high-fidelity (HF) models, while the low-fidelity (LF) models do not exist. Since VFS-HS involves both HF and LF function evaluations, modifications to the HF models are necessary to construct the LF models. The construction of LF models should adhere to the following two fundamental principles: (1) Ensure a reasonable difference between the HF and LF models, such that local exploration based on the LF model cannot obtain the global optimum of the HF model; (2) Maintain the general shape of the HF model's landscape in the LF model. Based on these principles, the HF and LF models used in VFS-HS are defined as follows:

$$f_H = f(x), \tag{8.5}$$

$$f_L = f((ax^T)^T), \tag{8.6}$$

where a is a defined translation matrix:

$$a = \begin{bmatrix} a_{11} & & & & \\ & \ddots & & & \\ & & a_{ii} & & \\ & & & \ddots & \\ & & & & a_{dd} \end{bmatrix}_{d\times d} \tag{8.7}$$

where a_{ii} is a random number within the range $[a_l, a_u]$. In this experiment, a_l and a_u are set to 0.9 and 1.1, respectively.

8.2.2 *Experimental Design and Parameter Settings*

The VFS-HS method proposed in this paper is compared with five existing surrogate-assisted optimization methods, namely the Particle Swarm Optimization algorithm with RBF surrogate (OPUS-RBF) [15], the Two-Layer Surrogate-Assisted Particle Swarm Optimization algorithm (TLSAPSO) [16], the Surrogate-Assisted Hierarchical Particle Swarm Optimization (SHPSO) [17], the Hybrid Surrogate Optimization method based on Space Reduction (HSOSR) [18], and the Memetic Differential Evolution assisted by Multi-fidelity Gaussian Process and Radial Basis Function

Table 8.3 Parameter settings of comparison algorithms

Algorithm	Parameters
OPUS-RBF	Swarm size = $2d$, $c_1 = c_2 = 1.496172$, $\omega = 0.72984$, side length of local region $\xi = 0.1 \cdot \min(UB - LB)$, trial positions for each particle $r = 10d$, threshold distance $\delta = 0.0005\sqrt{d} \cdot \min(UB - LB)$
TLSAPSO	Swarm size= $2d$, $c_1 = c_2 = 2.05$, $\phi = 4.1$, $K = 0.15$, maximum velocity $V_{max} = \min(UB - LB)/4$, spread coefficient $\alpha = 0.25$, threshold value for determining whether a new sampling point should be added to the global database or local database $\delta_1 = \delta_2 = 0.001$
SHPSO	Swarm size= 50, constriction factor $\chi = 0.729$, $c_1 = c_2 = 2.05$, size of Latin hypercube sampling $n_s = 100$, size of data to train RBF surrogate $n_t = 100$
HSOSR	Number of cheap points generated by LHS $n_c = 1000$, $n_{rank1} = 100$, $n_{rank2} = 50$, coefficient of distance $\Delta_1 = 10^{-3}$, $\Delta_2 = 10^{-6}$
MGPMDE	Threshold values of normalized population diversity $\delta_1 = 0.03$, $\delta_2 = 0.06$, number of groups in the pool $G = 5$, number of samples to represent each group $n_0 = 2$, size of seed population $\lambda_0 = 10$, population size of differential evolution $\lambda = 50$

models (MGPMDE) [19]. The first four methods belong to single-fidelity surrogate-assisted optimization methods, while the last one is a multi-fidelity surrogate-assisted optimization method. Table 8.3 presents the parameter settings related to the aforementioned comparison methods, all of which follow the suggestions provided in the original papers. All methods were implemented on a laptop with an Intel (R) Core (TM) i5-2540M CPU@2.6 GHz using MATLAB R2016b. For each method, 20 independent runs were conducted on all benchmark functions. For ease of comparison, it is assumed here that the computational time of 10 low-fidelity (LF) evaluations is equivalent to that of one high-fidelity (HF) evaluation, a common equivalence in industry [19, 20]. Since VFS-HS and MGPMDE involve both HF and LF evaluations, while SHPSO, HSOSR, OPUS-RBF, and TLSAPSO only involve HF evaluations, to ensure fairness in comparison, the computational budget for all methods was limited to within 50d times the equivalent real function evaluations.

In VFS-HS, several parameters need to be determined, namely HMS, $HMCR$, PAR, M, F, the local search radius R, and the maximum number of iterations for local search $maxLS$. Most of these parameters are derived from MHS, and we have previously studied their impact on the performance of MHS [9]. Based on experience, we followed most of the parameter settings from the previous study ($HMCR = 0.99$, $PAR = 0.90$, $M = 0.4$, $F = 0.5$), except for HMS. HMS represents the harmony memory size, which controls the population diversity: a larger HMS will help form a more diverse population but will slow down the convergence process. Typically, for problems with inexpensive function evaluations, we set $HMS = 100$. However, for the expensive optimization problems in this study, due to the limited maximum number of function evaluations, setting a large HMS is not appropriate. To balance population diversity and convergence speed, we set HMS to be a variable during the

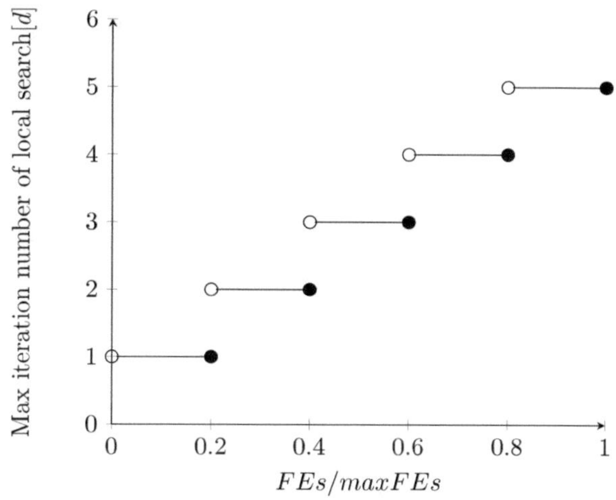

Fig. 8.5 Illustration of the step function for adjusting the maximum number of iterations for local search

optimization process. HMS is initially set to the size of the initial HF samples (set to $2d$ for the initial HF sample size and $10d$ for the initial LF sample size), and it is gradually increased in subsequent iterations until it reaches the maximum HMS ($maxHMS = \max(2d, 50)$). It is worth noting that for high-dimensional problems ($d \geq 25$), HMS remains constant during the optimization process. For the local search radius R, we set it to 0.1, the same as in OPUS-RBF.

Figure 8.5 illustrates the step function method used to adjust the maximum number of iterations for local search ($maxLS$) in VFS-HS. In the initial stage, the maximum number of iterations is set to a small value and gradually increased as the optimization progresses. This dynamic setting of the maximum number of iterations is based on two reasons. First, exploration and exploitation are the two cornerstones of meta-heuristic algorithms. Exploration refers to the ability to search for the global optimum in a diverse search space, while exploitation refers to the ability to fully utilize the favorable information from a previously promising solution to obtain a better solution. Therefore, in the early stages of optimization, we prefer exploration behavior, while in the later iterations, we aim to strengthen exploitation behavior. Second, in most surrogate-assisted optimization methods, the accuracy of the surrogate model is gradually improved during the optimization process. In the initial stage, local search should not be emphasized, as an inaccurate surrogate model may lead to unnecessary computational costs and sometimes even mislead the search direction.

8.2.3 Performance Comparison and Analysis

(a) Computational Results

Tables 8.4 and 8.5 list the statistical metrics (best, worst, median, mean, and standard deviation) for the 10-dimensional and 30-dimensional benchmark functions. For

Table 8.4 Final statistical results of all methods on benchmark functions (d = 10)

Func	Metrics	HSOSR	SHPSO	MGPMDE	OPUS-RBF	TLSAPSO	VFS-HS
F1	Best	2.44E−4	2.90E−5	**2.25E–6**	1.85E−4	1.58E−1	3.39E−6
	Worst	1.06E−1	2.24E−3	9.04E−5	1.22E+2	9.83E+0	**4.18E–5**
	Median	1.08E−2	1.17E−4	1.11E−5	3.86E+1	7.97E−1	**1.07E–5**
	Mean	1.81E−2	3.22E−4	1.68E−5	4.42E+1	2.71E+0	**1.53E–5**
	Std	2.35E−2	5.62E−4	1.96E−5	3.27E+1	3.46E+0	**1.06E–5**
F2	Best	5.03E−2	1.24E−3	**4.94E–5**	4.37E−2	4.86E−1	6.21E−5
	Worst	2.74E+1	6.71E−1	**1.59E–3**	2.14E+2	2.61E+1	1.75E−3
	Median	8.01E+0	2.94E−2	**2.48E–4**	7.07E+1	2.50E+0	3.78E−4
	Mean	9.29E+0	6.99E−2	**3.51E–4**	8.79E+1	4.63E+0	4.38E−4
	Std	8.35E+0	1.44E−1	3.64E−4	6.20E+1	6.58E+0	**3.45E–4**
F3	Best	1.07E+1	4.46E−1	**2.62E–4**	1.41E+0	1.74E+0	1.10E−3
	Worst	3.67E+2	7.52E+1	2.31E+2	3.62E+2	1.04E+2	**2.06E–2**
	Median	1.01E+2	6.36E+0	**1.55E–3**	1.13E+2	2.17E+1	4.85E−3
	Mean	1.37E+2	1.42E+1	2.29E+1	1.23E+2	3.29E+1	**6.25E–3**
	Std	1.01E+2	1.85E+1	5.43E+1	9.58E+1	3.21E+1	**4.92E–3**
F4	Best	0.00E+1	0.00E+1	5.00E+0	0.00E+1	1.00E+0	**0.00E+1**
	Worst	8.00E+0	1.00E+0	2.90E+0	2.14E+2	1.00E+1	**0.00E+1**
	Median	2.00E+0	0.00E+1	1.35E+1	4.50E+1	2.00E+0	**0.00E+1**
	Mean	2.45E+0	5.00E−2	1.40E+1	6.12E+1	3.05E+0	**0.00E+1**
	Std	2.48E+0	2.24E−1	6.13E+0	5.47E+1	2.44E+0	**0.00E+1**
F5	Best	2.66E+0	3.05E+0	5.29E+0	3.01E+0	**1.78E+0**	2.68E+0
	Worst	1.23E+1	9.06E+0	1.32E+1	9.45E+0	6.00E+0	**4.43E+0**
	Median	5.13E+0	5.79E+0	7.77E+0	6.09E+0	3.84E+0	**3.49E+0**
	Mean	5.78E+0	5.82E+0	8.35E+0	6.37E+0	3.78E+0	**3.49E+0**
	Std	2.61E+0	1.43E+0	2.18E+0	1.99E+0	1.29E+0	**5.36E–1**
F6	Best	6.98E−1	**3.28E–1**	4.63E−1	3.73E−1	6.97E−1	5.24E−1
	Worst	1.03E+0	**9.32E–1**	3.01E+0	2.02E+0	1.78E+0	1.02E+0
	Median	8.83E−1	**7.35E–1**	1.30E+0	1.00E+0	1.02E+0	8.15E−1
	Mean	8.77E−1	**7.38E–1**	1.58E+0	1.04E+0	1.07E+0	7.98E−1
	Std	**1.10E–1**	1.41E−1	8.40E−1	4.00E−1	2.57E−1	1.18E−1
F7	Best	6.23E+1	8.59E+0	9.61E+0	4.23E+1	**3.45E+0**	4.76E+0
	Worst	4.93E+2	2.02E+2	7.20E+1	6.53E+2	1.52E+2	**5.32E+1**
	Median	2.24E+2	2.83E+1	2.15E+1	1.73E+2	4.70E+1	**1.66E+1**
	Mean	2.57E+2	4.15E+1	2.64E+1	2.06E+2	5.30E+1	**1.85E+1**
	Std	1.25E+2	4.76E+1	1.67E+1	1.57E+2	4.49E+1	**1.06E+1**
F8	Best	3.74E+1	2.68E+1	7.96E+0	1.73E+1	1.63E+1	**6.12E+0**
	Worst	9.80E+1	6.86E+1	6.57E+1	**5.49E+1**	6.58E+1	7.30E+1
	Median	7.15E+1	5.08E+1	**1.94E+1**	3.38E+1	3.33E+1	5.05E+1
	Mean	7.02E+1	4.99E+1	**2.23E+1**	3.50E+1	3.65E+1	4.63E+1
	Std	1.81E+1	1.18E+1	1.34E+1	**8.87E+0**	1.38E+1	1.84E+1
Total scores		1.25	4.75	9	2.25	2	**20.75**

Table 8.5 Final statistical results of all methods on benchmark functions (d = 30)

Func	Metrics	HSOSR	SHPSO	MGPMDE	OPUS-RBF	TLSAPSO	VFS-HS
F1	Best	3.88E−1	3.80E+1	**2.21E–5**	2.67E+2	3.66E+1	4.76E−4
	Worst	2.03E+1	4.08E+2	**1.22E–4**	7.11E+2	2.10E+2	1.57E−3
	Median	3.15E+0	1.94E+2	**5.00E–5**	4.87E+2	1.01E+2	7.68E−4
	Mean	5.57E+0	1.89E+2	**5.45E–5**	4.76E+2	1.09E+2	8.69E−4
	Std	5.01E+0	8.62E+1	**2.49E–5**	1.31E+2	4.23E+1	3.38E−4
F2	Best	3.33E+1	1.29E+3	**7.93E–4**	2.59E+3	6.41E+2	1.92E−1
	Worst	2.14E+2	6.03E+3	**1.14E–2**	8.54E+3	2.75E+3	1.35E+0
	Median	1.15E+2	2.43E+3	**1.98E–3**	5.27E+3	1.73E+3	3.51E−1
	Mean	1.12E+2	2.55E+3	**2.51E–3**	5.49E+3	1.77E+3	4.62E−1
	Std	6.02E+1	1.05E+3	**2.31E–3**	1.59E+3	6.17E+2	3.13E−1
F3	Best	2.54E+1	1.88E+3	**6.91E–3**	4.24E+3	1.23E+3	1.68E+0
	Worst	2.02E+2	7.27E+3	1.89E+4	1.07E+4	6.40E+3	**2.70E+1**
	Median	4.76E+1	3.30E+3	1.10E+4	7.86E+3	3.24E+3	**6.54E+0**
	Mean	6.58E+1	3.91E+3	9.88E+3	7.55E+3	3.33E+3	**7.59E+0**
	Std	4.50E+1	1.68E+3	4.65E+3	1.81E+3	1.49E+3	**6.32E+0**
F4	Best	7.00E+0	8.60E+1	1.77E+2	2.00E+2	3.20E+1	**0.00E+1**
	Worst	4.60E+1	6.91E+2	5.05E+2	5.16E+2	1.76E+2	**0.00E+1**
	Median	1.85E+1	2.55E+2	3.74E+2	3.48E+2	1.06E+2	**0.00E+1**
	Mean	2.22E+1	2.81E+2	3.77E+2	3.62E+2	1.04E+2	**0.00E+1**
	Std	1.25E+1	1.43E+2	8.11E+1	8.45E+1	3.68E+1	**0.00E+1**
F5	Best	3.03E+0	9.63E+0	1.38E+1	8.28E+0	6.23E+0	**8.05E–1**
	Worst	4.86E+0	1.48E+1	1.86E+1	1.25E+1	1.13E+1	**2.29E+0**
	Median	3.56E+0	1.20E+1	1.56E+1	1.05E+1	8.79E+0	**1.51E+0**
	Mean	3.65E+0	1.20E+1	1.57E+1	1.07E+1	8.67E+0	**1.51E+0**
	Std	4.52E−1	1.36E+0	1.28E+0	9.70E−1	1.30E+0	**4.70E–1**
F6	Best	3.66E−1	1.08E+0	9.36E−1	2.02E+0	4.03E+0	**5.39E–1**
	Worst	1.00E+0	7.29E+0	8.16E+1	3.94E+0	1.96E+1	**9.82E–1**
	Median	5.15E−1	2.58E+0	2.40E+1	2.68E+0	7.24E+0	7.04E–1
	Mean	5.56E−1	3.10E+0	3.38E+1	2.75E+0	8.47E+0	6.98E−1
	Std	1.70E−1	1.96E+0	3.27E+1	4.96E−1	3.91E+0	**8.90E–2**
F7	Best	2.90E+2	1.71E+2	3.04E+2	7.79E+2	2.25E+2	**4.30E+1**
	Worst	2.23E+3	7.99E+2	2.32E+3	3.67E+3	8.28E+2	**2.22E+2**
	Median	5.29E+2	4.80E+2	8.47E+2	1.67E+3	3.87E+2	**1.10E+2**
	Mean	6.28E+2	5.03E+2	1.05E+3	1.84E+3	4.28E+2	**1.17E+2**
	Std	4.24E+2	1.48E+2	4.85E+2	7.93E+2	1.36E+2	**4.42E+1**
F8	Best	2.02E+2	1.08E+2	7.66E+1	1.43E+2	1.04E+2	**5.07E+1**
	Worst	3.89E+2	2.48E+2	2.31E+2	**2.28E+2**	2.57E+2	2.67E+2
	Median	3.06E+2	1.90E+2	**1.10E+2**	1.95E+2	2.02E+2	1.87E+2
	Mean	3.10E+2	1.85E+2	**1.19E+2**	1.92E+2	2.02E+2	1.53E+2
	Std	3.90E+1	3.11E+1	3.67E+1	**2.16E+1**	3.61E+1	8.31E+1
Total scores		2	0	13	2	0	**23**

each problem and each metric in Tables 8.4 and 8.5, the best values among different methods are highlighted in bold.

The algorithm has a total of 8 functions, each with 5 metrics (best, worst, median, mean, and standard deviation), resulting in a maximum possible score of 40 points. If an algorithm wins in a particular metric for a function, it receives 1 point; otherwise, it receives 0 points. If there are n winners in a competition, they share 1 point equally, each receiving $1/n$ points.

The results in Table 8.4 show that VFS-HS outperforms other methods on the 10-dimensional problems. Among the 6 competitors, VFS-HS achieved the highest score (20.75 points), followed by MGPMDE with 9 points. The worst-performing method was HSOSR, which only scored 1.25 points.

The results in Table 8.5 indicate that the performance of all methods deteriorated when the search space expanded from 10 dimensions to 30 dimensions. However, VFS-HS still outperformed the other methods. Similarly, among the 6 competitors, VFS-HS achieved the highest score (23 points), followed by MGPMDE with 13 points. The worst-performing algorithms were SHPSO and TLSAPSO, which did not win in any of the competitions.

(b) Wilcoxon Signed-Rank Test

Due to the stochastic nature of metaheuristic algorithms such as PSO and HS, statistical tests are necessary. In this study, the Wilcoxon signed-rank test [21, 22] was employed to detect significant differences between VFS-HS and its competitors (if any). The significance level was set to $\alpha = 0.05$. If the p-value is less than or equal to 0.05, the null hypothesis that "there is no significant difference between VFS-HS and its competitor" is rejected; otherwise, the hypothesis is accepted. The results of the Wilcoxon signed-rank test are presented in Tables 8.6 and 8.7. The symbols "+" and "−" in the tables indicate that VFS-HS significantly outperformed or underperformed its competitor, respectively, while "=" indicates no significant difference. R+ represents the total number of ranks where VFS-HS outperformed the comparison method, and R− represents the opposite. The results in Tables 8.6 and 8.7 show that VFS-HS has more "+" counts than "=" and "−" counts, indicating its statistical superiority over other methods.

(c) Convergence Process

To investigate the convergence characteristics of all compared methods, Figs. 8.6 and 8.7 display the average convergence curves of the surrogate methods on the 10-dimensional and 30-dimensional benchmark functions, respectively.

For the 10-dimensional problems, VFS-HS demonstrated superior convergence performance on F_3, F_4, F_5, and F_7 compared to all other methods. For F_1 and F_2, the convergence performance of VFS-HS was slightly inferior to that of MGPMDE but superior to other methods. A similar situation was observed for F_6, where VFS-HS's convergence performance was slightly inferior to SHPSO but better than other methods. For F_8, VFS-HS ranked fourth in terms of convergence performance

Table 8.6 Wilcoxon signed-rank test results based on the best objective function values (each benchmark function run independently 20 times, $d = 10$, significance level $\alpha = 0.05$)

Function	VFS-HS versus HSOSR				VFS-HS versus SHPSO				VFS-HS versus MGPMDE			
	R+	R−	p-value	Win	R+	R−	p-value	Win	R+	R−	p-value	Win
F_1	210.0	0.0	8.86E−5	+	209.0	1.0	1.03E−4	+	105.0	105.0	1.00E+0	=
F_2	210.0	0.0	8.86E−5	+	210.0	0.0	8.86E−5	+	76.0	134.0	2.79E−1	=
F_3	210.0	0.0	8.86E−5	+	210.0	0.0	8.86E−5	+	96.0	114.0	7.37E−1	=
F_4	210.0	0.0	9.45E−4	+	0.0	210.0	3.17E−1	=	210.0	0.0	8.76E−5	+
F_5	195.0	15.0	7.80E−4	+	207.0	3.0	1.40E−4	+	210.0	0.0	8.86E−5	+
F_6	150.0	60.0	9.30E−2	=	71.0	139.0	2.04E−1	=	195.0	15.0	7.80E−4	+
F_7	210.0	0.0	8.86E−5	+	163.0	47.0	3.04E−2	+	135.0	75.0	2.63E−1	=
F_8	187.0	23.0	2.20E−3	+	126.0	84.0	4.33E−1	=	11.0	199.0	4.49E−4	−
Function	VFS-HS versus OPUS-RBF				VFS-HS versus TLSAPSO							
	R+	R−	p-value	Win	R+	R−	p-value	Win				
F_1	210.0	0.0	8.86E−5	+	210.0	0.0	8.86E−5	+				
F_2	210.0	0.0	8.86E−5	+	210.0	0.0	8.86E−5	+				
F_3	210.0	0.0	8.86E−5	+	210.0	0.0	8.86E−5	+				
F_4	210.0	0.0	1.96E−4	+	210.0	0.0	8.00E−5	+				
F_5	206.0	4.0	1.63E−4	+	130.0	80.0	3.51E−1	=				
F_6	171.0	39.0	1.37E−2	+	203.0	7.0	2.54E−4	+				
F_7	210.0	0.0	8.86E−5	+	180.0	30.0	5.11E−3	+				
F_8	50.0	160.0	4.00E−2	−	54.0	156.0	5.69E−2	=				

among all compared methods. For the 30-dimensional problems, the advantages of VFS-HS were more pronounced. Similarly, VFS-HS outperformed all other methods in terms of convergence on F_3, F_4, F_5, and F_7. For F_1, F_2, and F_8, the convergence performance of VFS-HS was slightly inferior to that of MGPMDE but better than other methods. For F_6, VFS-HS's convergence performance was slightly inferior to HSOSR but better than other methods.

(d) Computational Time Analysis

To analyze the computational time of VFS-HS, Table 8.8 lists the average computational time of all comparison methods on the benchmark functions. Among the six comparison methods, HSOSR is the slowest. This is because HSOSR calls multi-start local search in each iteration, which is very expensive. The average computational cost of VFS-HS is similar to that of MGPMDE for 10-dimensional problems, but it becomes slower for 30-dimensional problems. SHPSO, OPUS-RBF, and TLSAPSO run faster than other methods. Overall, the average computational time of VFS-HS is acceptable. In practical applications, a single function evaluation may take anywhere from several minutes to several hours, so compared with these expensive function evaluations, the computational cost of VFS-HS can be negligible. For example, assuming that a high-fidelity (HF) evaluation of problem F_8 (30-dimensional) takes 2 min, the non-evaluation time of VFS-HS accounts for less than 0.3% of the total

Table 8.7 Wilcoxon signed-rank test results based on the best objective function values (each benchmark function run independently 20 times, $d = 30$, significance level $\alpha = 0.05$)

Function	VFS-HS versus HSOSR				VFS-HS versus SHPSO				VFS-HS versus MGPMDE			
	R+	R−	*p*-value	Win	R+	R−	*p*-value	win	R+	R−	*p*-value	Win
F_1	210.0	0.0	8.86E−5	+	210.0	0.0	8.86E−5	+	0.0	210.0	8.86E−5	−
F_2	210.0	0.0	8.86E−5	+	210.0	0.0	8.86E−5	+	0.0	210.0	8.86E−5	−
F_3	210.0	0.0	8.86E−5	+	210.0	0.0	8.86E−5	+	207.0	3.0	1.40E−4	+
F_4	210.0	0.0	8.78E−5	+	210.0	0.0	8.86E−5	+	210.0	0.0	8.81E−5	+
F_5	210.0	0.0	8.86E−5	+	210.0	0.0	8.86E−5	+	210.0	0.0	8.86E−5	+
F_6	35.0	175.0	8.97E−3	−	210.0	0.0	8.86E−5	+	210.0	0.0	8.86E−5	+
F_7	210.0	0.0	8.86E−5	+	210.0	0.0	8.86E−5	+	210.0	0.0	8.86E−5	+
F_8	209.0	1.0	1.03E−4	+	67.0	143.0	1.56E−1	=	60.0	150.0	9.30E−2	=
Function	VFS-HS versus OPUS-RBF				VFS-HS versus TLSAPSO							
	R+	R−	*p*-value	Win	R+	R−	*p*-value	Win				
F_1	210.0	0.0	8.86E−5	+	210.0	0.0	8.86E−5	+				
F_2	210.0	0.0	8.86E−5	+	210.0	0.0	8.86E−5	+				
F_3	210.0	0.0	8.86E−5	+	210.0	0.0	8.86E−5	+				
F_4	210.0	0.0	8.86E−5	+	210.0	0.0	8.81E−5	+				
F_5	210.0	0.0	8.86E−5	+	210.0	0.0	8.86E−5	+				
F_6	210.0	0.0	8.86E−5	+	210.0	0.0	8.86E−5	+				
F_7	210.0	0.0	8.86E−5	+	210.0	0.0	8.86E−5	+				
F_8	151.0	59.0	8.59E−2	=	169.0	41.0	1.69E−2	+				

computational time. Even for HSOSR, the non-evaluation time only accounts for 2.7% of the total computational time. Under such circumstances, the difference in non-evaluation time between VFS-HS and other methods has a very limited impact on the total computational time.

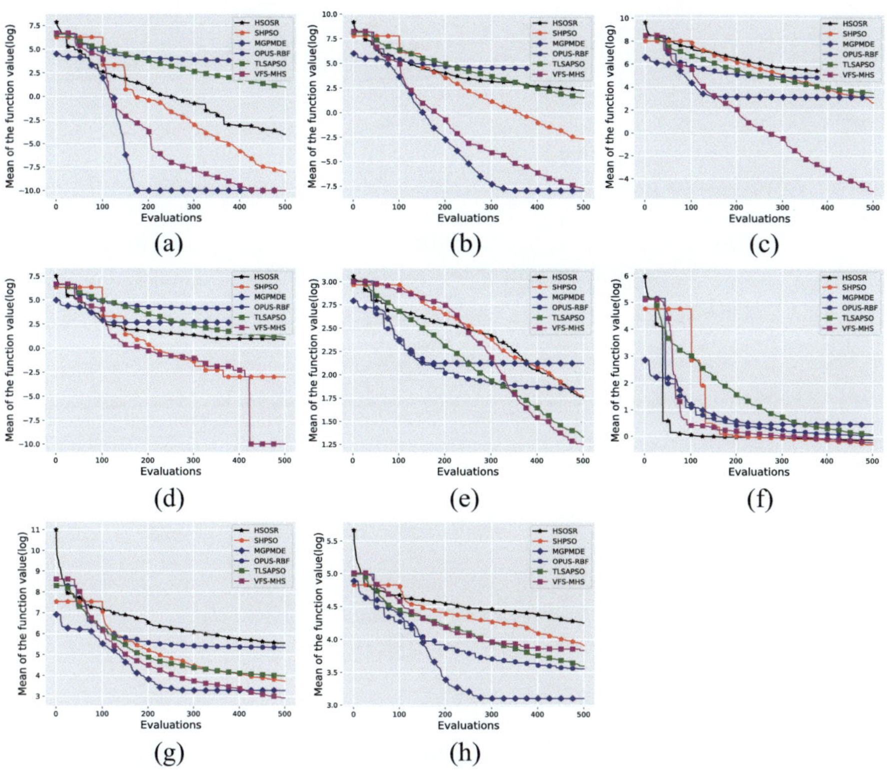

Fig. 8.6 Convergence curves of alternative optimization methods on 10-dimensional benchmark functions (F1–F8)

(e) Robustness Analysis of VFS-HS

In practical engineering optimization problems, the degree of difference between the HF model and the LF model is problem-dependent, sometimes small and sometimes large. Therefore, variable-fidelity surrogate-assisted optimization methods should have stable performance (high robustness) when dealing with problems involving different model gaps. To investigate the robustness of VFS-HS, its performance was further tested in this section. In Sect. 8.2.1, we defined the shift matrix $a = \{a_{ii}\}$, $i = 1, 2, \ldots, d$, where $a = \{a_{ii}\}$ is a random number within the range $[a_l, a_u]$.

Here, we introduce a new parameter θ, and set $a_l = 1.0 - \theta$ and $a_u = 1.0 + \theta$. The parameter θ represents the degree of difference between the HF model and the LF model: a larger value indicates a greater degree of difference.

In this experiment, we set θ to 0.1, 0.2, 0.3, 0.4, and 0.5, and ran VFS-HS on the same test problems as in Table 8.2 to determine whether there is a difference in performance. As in the previous experiment, VFS-HS was run 20 times for each problem. Figures 8.8 and 8.9 show the box plots of the results of 20 runs of VFS-HS at

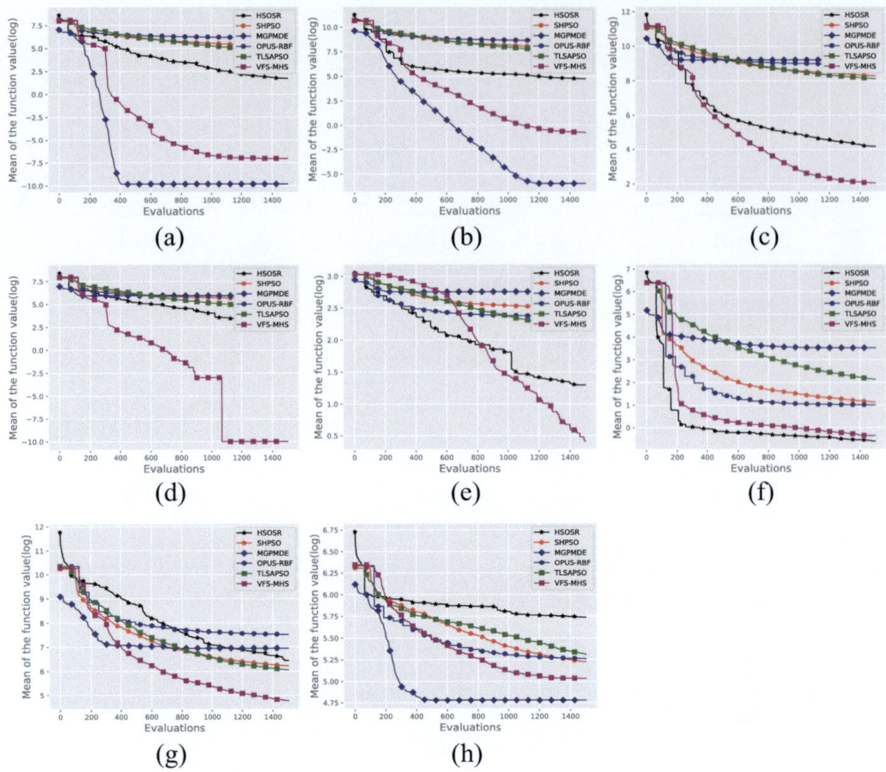

Fig. 8.7 Convergence curves of alternative optimization methods on 30-dimensional benchmark functions (F1–F8)

different values of θ. In the box plots, the maximum value, minimum value, median, first quartile, and third quartile are shown, with each point's position indicating the value of the corresponding run. It is worth noting that the box plot for F_4 is not provided, as VFS-HS obtained the global optimal solution (0) in all cases. From Figs. 8.8 and 8.9, it can be seen that there is no obvious trend indicating that a larger value of θ leads to worse performance (or vice versa), and the performance is also problem-dependent.

To determine whether there is a significant difference in the performance of VFS-HS at different values of θ, the Friedman test was employed. The null hypothesis is that VFS-HS with different values of θ has similar performance. The significance level was set to $\alpha = 0.05$ to determine whether to reject the hypothesis in all comparisons. Table 8.9 shows the results of the Friedman test. Although the Friedman rank value for $\theta = 0.1$ (equal to 2.3125) is smaller than the others, the p-value calculated from the Friedman statistic is $p = 0.1314$, which strongly indicates that there is no significant difference in the performance of VFS-HS at different values of θ.

Table 8.8 Average computational time of all methods on test benchmark functions (unit: s)

Dimensions	Functions	HSOSR	SHPSO	MGPMDE	OPUS-RBF	TLSAPSO	VFS-HS
10-d	F_1	1564.76	19.60	12.25	8.78	1.13	14.04
	F_2	1701.09	6.13	13.12	8.76	1.16	15.14
	F_3	1781.97	5.49	15.81	9.18	1.17	14.41
	F_4	1698.76	6.09	13.47	8.86	1.48	13.70
	F_5	1773.81	3.67	19.39	8.76	1.13	63.47
	F_6	1730.02	6.05	13.93	8.76	1.31	30.84
	F_7	1797.28	5.12	50.26	8.72	1.22	27.39
	F_8	1867.87	3.90	44.13	8.82	1.75	37.39
30-d	F_1	3101.66	11.50	95.82	94.02	7.42	123.69
	F_2	2785.93	11.18	144.77	93.98	7.92	136.19
	F_3	2534.78	11.05	174.27	94.04	7.97	143.37
	F_4	5133.92	11.17	83.92	93.98	8.49	149.69
	F_5	4898.43	10.74	104.09	94.00	8.23	294.42
	F_6	4990.21	13.84	134.21	93.98	7.80	331.80
	F_7	4744.23	10.95	246.08	93.95	8.48	463.77
	F_8	4902.36	11.05	190.04	94.04	14.63	476.84

Table 8.9 Results of statistical tests (significance level $\alpha = 0.05$)

θ	Friedman's test		
	Ranks	Friedman statistic	p-value
$\theta = 0.1$	2.3125	7.0857	0.1314
$\theta = 0.2$	2.9375		
$\theta = 0.3$	2.8125		
$\theta = 0.4$	3.5625		
$\theta = 0.5$	3.3750		

Therefore, it can be concluded that VFS-HS has good robustness for problems with different degrees of model bias.

8.2.4 Practical Application of Expensive Design Optimization Based on VFS-HS Algorithm: A Case Study of Pressure Vessel Optimization Design

To demonstrate the capability of the proposed VFS-HS in solving expensive real-world engineering design problems, this section investigates the example of a long cylindrical high-pressure container designed for Compressed Natural Gas

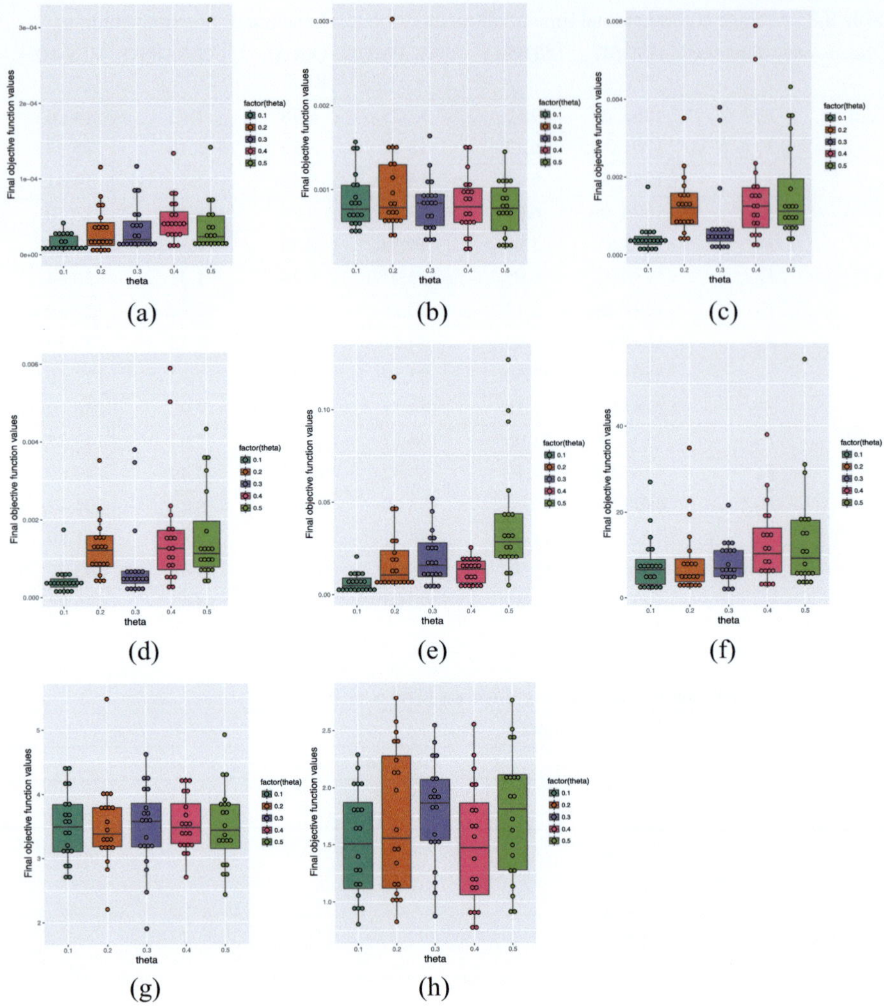

Fig. 8.8 Box plots of final objective function values at different model difference levels (F1–F3, F5)

(CNG) [23, 24]. The outer diameter and length of the container's main body are 406 and 6000 mm, respectively. As shown in Fig. 8.10, the container is subjected to a uniform pressure of 23 MPa. The material's elastic modulus and Poisson's ratio are $E = 207$ GPa and $\mu = 0.30$, respectively. To enhance the strength of the end regions, the thickness in these areas gradually increases from the main body. The design objective is to minimize the stress while ensuring that the chamber volume is not less than $0.63\,\text{m}^3$ and the maximum material consumption for manufacturing does not exceed $0.12\,\text{m}^3$. The design variables are shown in Table 8.10. The mathematical model of this problem can be expressed as follows:

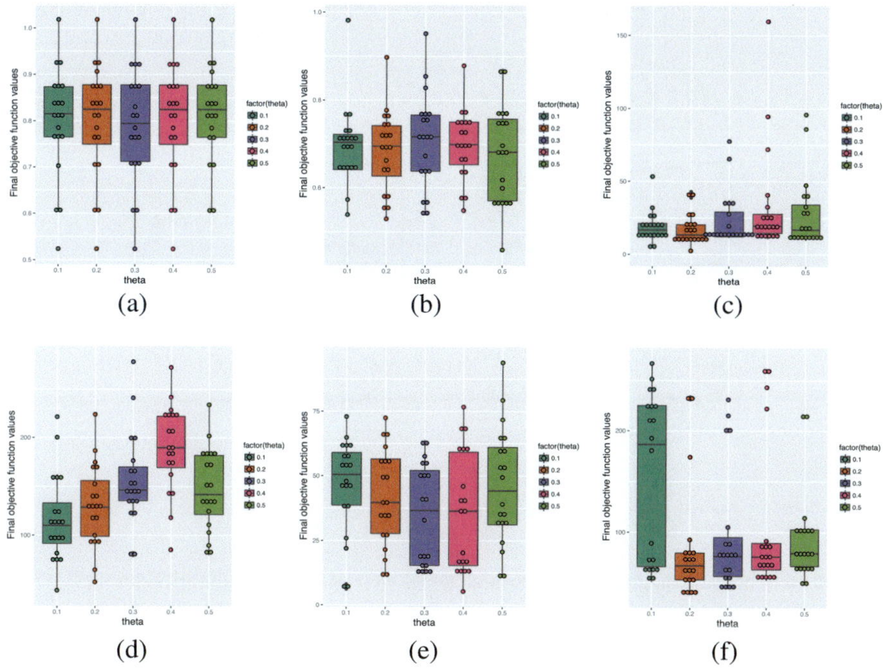

Fig. 8.9 Box plots of final objective function values at different model difference levels (F6–F8)

$$
\begin{aligned}
&\text{Minimize:} \quad f(h_1, b_1, C_b, r_1, C_r) = \sigma_s, \\
&\text{s.t.} \quad \begin{cases} M \leq 0.12\,\text{m}^3 \\ V \geq 0.63\,\text{m}^3, \end{cases}
\end{aligned}
\tag{8.8}
$$

where σ_s is the maximum von Mises stress of the material, M is the material consumption for manufacturing, and V is the internal volume of the container. M and V can be calculated using the following equations:

$$
\begin{aligned}
\text{M} = &\pi[(C_r + r_1)^2 - r_1^2](6000 - h_1 - \sqrt{r_1^2 - b_1^2}) \\
&+2\pi h_1[(C_b + b_1)^2 - b_1^2] \\
&+\pi[(C_b + b_1)^2 + (C_r + r_1)^2]\sqrt{(C_r + r_1)^2 - (C_b + b_1)^2} \\
&+\tfrac{1}{3}\pi[\sqrt{(C_r + r_1)^2 - (C_b + b_1)^2}]^3 \\
&-\tfrac{1}{3}\pi[3(r_1^2 + b_1^2)\sqrt{r_1^2 - b_1^2} + (\sqrt{r_1^2 - b_1^2})^3],
\end{aligned}
\tag{8.9}
$$

$$
V = \pi r_1^2(6000 - h_1 - \sqrt{r_1^2 - b_1^2}) + 2\pi h_1 b_1^2 + \pi(b_1^2 + r_1^2)\sqrt{r_1^2 - b_1^2} + \frac{1}{3}(\sqrt{r_1^2 - b_1^2})^3. \tag{8.10}
$$

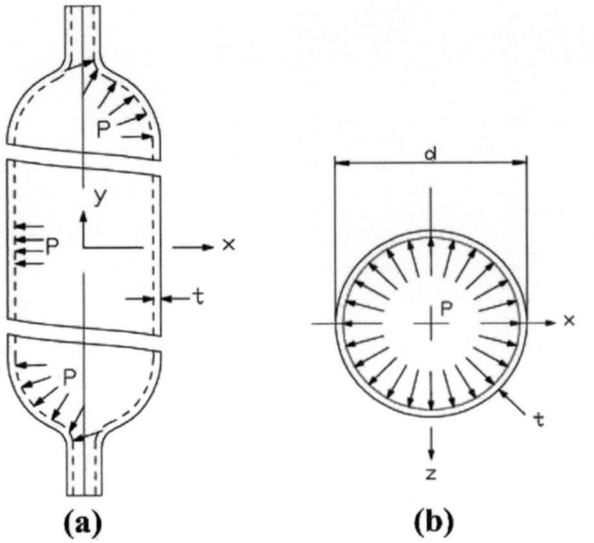

Fig. 8.10 **a** Schematic diagram and **b** cross-sectional view of the long cylindrical pressure vessel

Table 8.10 Ranges of design variables

Design variable	Range (mm)
End height h_1	280–320
End inner diameter b_1	25–60
End thickness C_b	15–30
Main body inner diameter r_1	150–230
Main body thickness C_r	10–23

However, the maximum von Mises stress only has an implicit expression. Therefore, a variable-fidelity surrogate model was developed to fit the relationship between the stress response and the design variables. In this study, the stress response was obtained using ANSYS 18.0. An axisymmetric three-dimensional finite element model with a 15 mm hexahedral mesh was selected as the high-fidelity (HF) model, as shown in Fig. 8.11. A three-dimensional finite element model with a 40 mm hexahedral mesh was used as the low-fidelity (LF) model, as shown in Fig. 8.12.

The problem was then solved using the improved VFS-HS method proposed in this paper, and its effectiveness was verified by comparing it with HSOSR, SHPSO, OPUS-RBF, TLSAPSO, and MGPMDE. Unlike the numerical experiments in Sect. 4, the optimization design problem of the cylindrical gas-pressure vessel also involves additional constraints. To handle these constraints, Deb's constraint handling method [25] was employed in this study. In the VFS-HS, the initial sampling sizes for the high-fidelity (HF) model and low-fidelity (LF) model were set to 10 and 50, respectively. The HF simulation took approximately 22.0 s, while the LF simulation

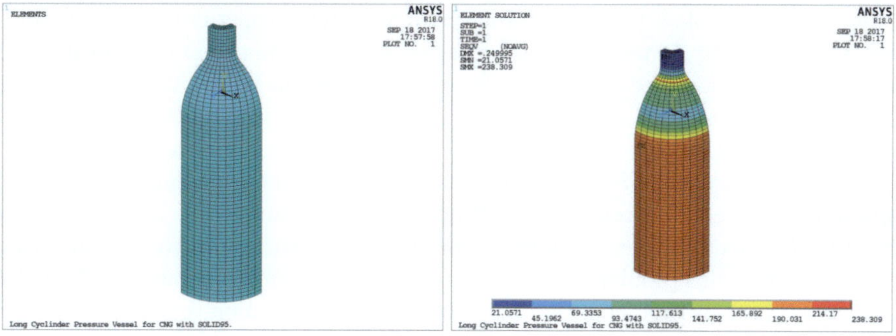

(a) (b)

Fig. 8.11 HF model of the cylindrical pressure vessel using a 15 mm hexahedral mesh

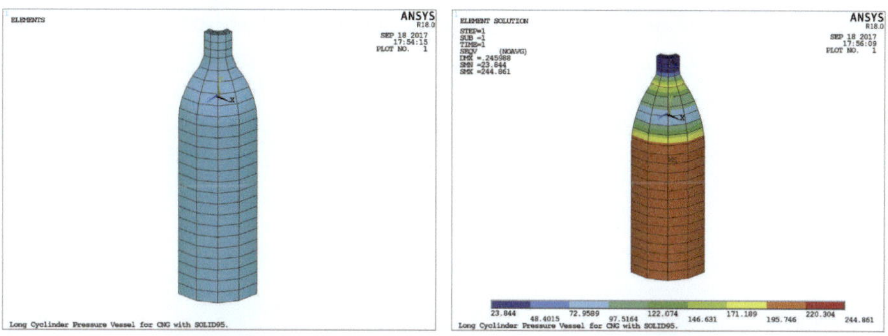

(a) Grid model (b) Simulation analysis

Fig. 8.12 LF of the cylindrical pressure vessel using a 40 mm hexahedral mesh

Table 8.11 Final results of all methods on the long cylindrical pressure vessel optimal design problem

Metric	HSOSR	SHPSO	MGPMDE	OPUS-RBF	TLSAPSO	VFS-HS
Best value	265.004	264.899	265.513	266.383	265.135	**264.845**
Worst value	271.066	274.692	275.939	273.613	274.587	**265.162**
Median	266.031	268.036	272.030	268.934	269.558	**264.967**
Mean	266.519	268.448	271.067	269.109	269.536	**264.977**
Standard deviation	1.607	2.718	3.643	1.989	2.633	**0.089**

took about 2.0 s. In this experiment, all algorithms were terminated when the number of "equivalent" HF evaluations reached 500. Table 8.11 presents the results obtained by all comparison methods over 20 trials.

As shown in Table 8.11, the performance of VFS-HS is superior to all other methods. The Wilcoxon signed-rank test was again employed to detect whether the performance differences between the methods are significant. If the calculated p-value is less than 0.05, the null hypothesis (that there is no significant difference between VFS-HS and its competitor) will be rejected. Similarly, the symbols "+" and "−" in the table indicate that VFS-HS significantly outperforms or underperforms its competitor, respectively. The results of the Wilcoxon signed-rank test in Table 8.12 show that VFS-HS obtained the "+" symbol for all five comparison methods, indicating its superiority in this problem. Table 8.13 presents the average computational time of all methods for this problem. The first column "Non-evaluation time" in Table 8.13 represents the computational time excluding the expensive optimization objective function evaluation time, while "Total computational time" represents the total computational time including both the optimizer and the expensive optimization objective function evaluation time. As shown in Table 8.13, although the non-evaluation times of all comparison methods vary from tens of seconds to hundreds of seconds, the

Table 8.12 Results of the Wilcoxon rank-sum test for the long cylindrical pressure vessel optimization design problem based on 20 independent runs (significance level $\alpha = 0.05$)

VFS-HS versus	R+	R−	p-value	win
HSOSR	210.0	0.0	8.86E−5	+
SHPSO	208.0	2.0	1.20E−4	+
MGPMDE	210.0	0.0	8.86E−5	+
OPUS-RBF	210.0	0.0	8.86E−5	+
TLSAPSO	210.0	0.0	8.86E−5	+

Table 8.13 Average computational time of all methods for the long cylindrical pressure vessel optimization design problem (unit: hours)

Metric	HSOSR	SHPSO	MGPMDE	OPUS-RBF	TLSAPSO	VFS-HS
Non-evaluation time	0.122	0.033	0.036	0.006	0.018	0.016
Total computational time	3.177	3.088	3.092	3.062	3.074	3.072

differences in total computational time are very small in scale (with the largest time difference being only 3.75%). Therefore, the computational costs of all comparison methods are quite reasonable.

8.3 Summary

In this research, an online variable-fidelity surrogate-assisted harmony search algorithm is proposed to address costly engineering design optimization challenges. The variable-fidelity surrogate model integrates data from various fidelity levels, enabling it to generate a highly precise approximation of the computationally intensive function, even when there are limited high-fidelity (HF) samples available. Additionally, the VFS-HS algorithm utilizes an innovative management approach that evenly distributes HF evaluations throughout the optimization procedure, thereby guiding the search process effectively. Furthermore, a multi-level screening mechanism based on non-dominated sorting is introduced to regulate the number of low-fidelity (LF) and HF evaluations, ensuring a proper balance between exploration and exploitation.

The proposed VFS-HS algorithm demonstrates its effectiveness and efficiency through a comparative analysis with four established single-fidelity surrogate-assisted methods (SHPSO, HSOSR, OPUS-RBF, and TLSAPSO) and one multi-fidelity surrogate-assisted optimization approach known as MGPMDE. The experimental outcomes indicate that VFS-HS surpasses these competing methods in most test cases regarding both solution quality and convergence rate. A robustness evaluation of VFS-HS reveals not only its stability but also its resilience to model inconsistencies. Moreover, a practical engineering application involving the design of a long cylindrical gas-pressure vessel highlights VFS-HS's ability to address real-world, computationally expensive engineering design optimization challenges. Given its superior performance compared to other methods, VFS-HS can be regarded as a promising technique for solving such optimization problems when computationally affordable low-fidelity (LF) models are accessible.

References

1. Geem ZW, Kim JH, Loganathan GV (2001) A new heuristic optimization algorithm: harmony search. Simulation 76(2):60–68
2. Kundu S, Parhi DR (2016) Navigation of underwater robot based on dynamically adaptive harmony search algorithm. Memetic Comput 8(2):125–146
3. Zeng W, Yi J, Rao X et al (2017) A two-stage path planning approach for multiple car-like robots based on PH curves and a modified harmony search algorithm. Eng Optim 49(11):1995–2012
4. Yi Jin, Chu Chih-Hsing, Kuo Chi-Lung et al (2018) Optimized tool path planning for five-axis flank milling of ruled surfaces using geometric decomposition strategy and multi-population harmony search algorithm. Appl Soft Comput 73:547–561
5. Maleki Akbar, Pourfayaz Fathollah (2015) Sizing of stand-alone photovoltaic/wind/diesel system with battery and fuel cell storage devices by harmony search algorithm. J Energy Storage 2:30–42
6. Zong Woo Geem and Yourim Yoon (2017) Harmony search optimization of renewable energy charging with energy storage system. Int J Electr Power Energy Syst 86:120–126
7. Chen J, Pan QK, Li JQ (2012) Harmony search algorithm with dynamic control parameters. Appl Math Comput 219(2):592–604
8. Yi Jin, Li Xinyu, Chu Chih-Hsing et al (2019) Parallel chaotic local search enhanced harmony search algorithm for engineering design optimization. J Intell Manuf 30:405–428
9. Yi Jin, Gao Liang, Li Xinyu et al (2016) An efficient modified harmony search algorithm with intersect mutation operator and cellular local search for continuous function optimization problems. Appl Intell 44:725–753
10. Črepinšek M, Liu SH, Mernik M (2013) Exploration and exploitation in evolutionary algorithms: a survey. ACM Comput Surv (CSUR) 45(3):1–33
11. Yi J, Li X, Xiao M et al (2017) Construction of nested maximin designs based on successive local enumeration and modified novel global harmony search algorithm. Eng Optim 49(1):161–180
12. Mehmani A, Chowdhury S, Tong W et al (2015) Adaptive switching of variable-fidelity models in population-based optimization.In: Engineering and applied sciences optimization: dedicated to the memory of professor MG Karlaftis, pp 175–205
13. Park C, Haftka RT, Kim NH (2017) Remarks on multi-fidelity surrogates. Struct Multidiscip Optim 55:1029–1050
14. Liu B, Chen Q, Zhang Q et al (2013) Problem definitions and evaluation criteria for computational expensive optimization. In: Technical report
15. Regis RG (2014) Particle swarm with radial basis function surrogates for expensive black-box optimization. J Comput Sci 5(1):12–23
16. Sun Chaoli, Jin Yaochu, Zeng Jianchao et al (2015) A two-layer surrogate-assisted particle swarm optimization algorithm. Soft Comput 19:1461–1475
17. Haibo Yu, Tan Ying, Zeng Jianchao et al (2018) Surrogate-assisted hierarchical particle swarm optimization. Inf Sci 454:59–72

18. Dong Huachao, Song Baowei, Wang Peng et al (2018) Hybrid surrogate-based optimization using space reduction (HSOSR) for expensive black-box functions. Appl Soft Comput 64:641–655
19. Liu Bo, Koziel Slawomir, Zhang Qingfu (2016) A multi-fidelity surrogate-model-assisted evolutionary algorithm for computationally expensive optimization problems. J Comput Sci 12:28–37
20. Koziel S, Leifsson L, Yang XS (2014) Solving computationally expensive engineering problems: methods and applications, vol 97. Springer, Berlin
21. Derrac J, García S, Molina D et al (2011) A practical tutorial on the use of nonparametric statistical tests as a methodology for comparing evolutionary and swarm intelligence algorithms. Swarm Evol Comput 1(1):3–18
22. Chao Lu, Gao Liang, Yi Jin (2018) Grey wolf optimizer with cellular topological structure. Expert Syst Appl 107:89–114
23. Zhou Qi, Wang Yan, Choi Seung-Kyum et al (2018) A robust optimization approach based on multi-fidelity metamodel. Struct Multidiscip Optim 57:775–797
24. Cai Xiwen, Qiu Haobo, Gao Liang et al (2017) Metamodeling for high dimensional design problems by multi-fidelity simulations. Struct Multidiscip Optim 56:151–166
25. Deb K (2000) An efficient constraint handling method for genetic algorithms. Comput Methods Appl Mech Eng 186(2–4):311–338

Chapter 9
Expensive Design Optimization With Transfer Learning-Based Sequential Variable-Fidelity Surrogate

Previous chapters introduced adaptive sequential variable-fidelity surrogate modeling, demonstrating its superiority. The efficacy of this approach hinges on selecting appropriate update strategies during the sequential process to complete modeling and optimization. Therefore, developing a universal infill strategy that leverages optimization information to enhance surrogate model accuracy while reducing computational costs is essential.

This chapter introduces an infill strategy based on the Non-Dominated Sorting Method (NDSM) and conducts sequential optimization based on the modeling method from Chap. 7, establishing a novel sequential optimization method. Its effectiveness is validated through relevant examples. This method effectively balances global exploration and local exploitation, enhancing model accuracy and optimization efficiency.

9.1 Sequence Optimization Method Based on Non-dominated Sorting Method

Engineering design optimization problems are typically formulated as

$$\begin{cases} \min \quad f(x, y(x)) \\ g_i(x, y(x)) \leq 0, \quad i = 1, \ldots, p, \\ h_j(x, y(x)) = 0, \quad j = 1, \ldots, q, \\ x_L \leq x \leq x_U. \end{cases} \tag{9.1}$$

Sequential optimization is an iterative optimization strategy. At each iteration step, design variables are updated based on previous design information and evaluation results to progressively approach the optimal solution. Unlike traditional one-time

J. Yi et al., *Variable-Fidelity Surrogate*, Engineering Applications of Computational Methods 25, https://doi.org/10.1007/978-981-95-5527-7_9

optimization design, sequential optimization fully utilizes information generated during the process, continuously adjusting the optimization direction. Its core lies in constructing an initial surrogate model with a small number of initial samples and then updating and optimizing the model by adding sample points. Updating samples according to the infill strategy is a crucial step in the sequential optimization process. It effectively controls the iterative update of the surrogate model and emphasizes balancing global exploration and local exploitation. Focusing solely on exploitation can cause evaluation points to cluster in a local region, leading to local optima. Focusing solely on exploration can result in insufficient sampling in certain regions, reducing local prediction accuracy.

Common strategies for balancing exploration and exploitation include the Expected Improvement (EI) strategy, which considers the relationship between exploration and exploitation but has drawbacks. It may not sufficiently explore regions with high prediction uncertainty. The EI strategy focuses more on local exploitation prediction, improving local prediction accuracy but suffering from lower global prediction accuracy and the risk of local optima. The Lower Confidence Bound (LCB) strategy considers both local exploitation and global exploration, enabling the selection of appropriate points for infill. However, unreasonable hyperparameter settings can shift the focus of the infill strategy, causing an imbalance between local exploitation and global exploration.

To address the limitations of traditional sequential optimization methods and effectively balance global exploration and local exploitation during sequential optimization, this chapter proposes a generalized infill strategy suitable for any model: the Non-dominated Sorting Method-based infill strategy (NDSM) and the corresponding sequence optimization method. The following sections detail sequential optimization based on NDSM, including the NDSM infill strategy and the NDSM-based sequential optimization method combined with the modeling method from Chap. 8.

9.1.1 Prediction Metrics

Model prediction metrics serve as indicators for evaluating global exploration and local exploitation. Global exploration pertains to exploring all areas of the variable space and can be assessed using distance-based metrics. The prediction accuracy of a surrogate model is closely related to the sparsity of the sample distribution. Sparse samples lead to larger prediction errors. Generally, the greater the sparsity, the larger the minimum distance between neighboring points; the denser the samples, the smaller this distance. Sample sparsity is defined based on the distance between samples, specifically the minimum distance from a target sample point to existing sample points:

$$d(x_i, x_j) = \sqrt{(x_i - x_j)^2} \tag{9.2}$$

$$D(x \mid S_t) = \min d(x, x_t), \tag{9.3}$$

where S_t represents the set of existing sample points and $D(x \mid S_t)$ represents the distance from a point x to the nearest existing sample point, i.e., the sample sparsity. Its geometric meaning is illustrated in Fig. 9.1.

The distance metric aims to select sample points that promote uniform distribution in the variable space, supplementing sparse regions and balancing sample distribution. Adding new samples updates the model, expressed as

$$\max D(x \mid S_t). \tag{9.4}$$

The distance metric can be effectively applied to global exploration for surrogate models, serving as an indicator for global exploration in the design space. For local exploitation, which focuses on searching near the region of the best predicted value by the surrogate model, the Minimum Surrogate Prediction (MSP) criterion can be used as the local exploitation metric. Its principle is to select sample points in the variable space that minimize the predicted value of the surrogate model to update the model. Adding these points refines the model:

$$\min \hat{f}(x), \tag{9.5}$$

where $\hat{f}(x)$ represents the surrogate model. This criterion emphasizes prediction accuracy, minimizing the predicted value, and is suitable for optimizing surrogate

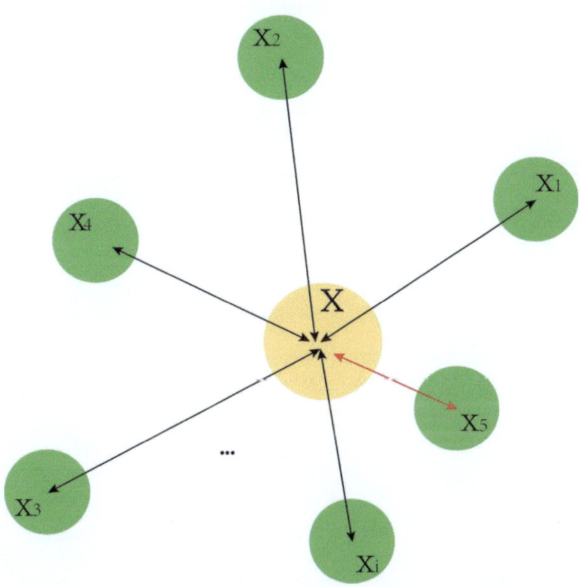

Fig. 9.1 Geometric interpretation of the minimum distance between sample points

models. This paper adopts the MSP criterion as the metric for local exploitation in the variable space.

Having established metrics for global exploration (distance metric) and local exploitation (MSP criterion) in the variable space, to maintain their balance during sequential optimization of the variable-fidelity surrogate model, they are treated as two conflicting objectives in the infill strategy:

$$\begin{cases} \max\limits_{x} \operatorname{mind}(x, x_t) \\ \min\limits_{x} \hat{f}(x) \\ x_L \leq x \leq x_U. \end{cases} \tag{9.6}$$

Since the optimization of the infill strategy involves two conflicting objectives, it is difficult to find a single optimal solution for both objectives. Solutions to multi-objective optimization problems typically form a set of solutions, known as Pareto optimal solutions. Within this set, no solution is superior to another in *all* objectives; these solutions are non-dominated and equally valid. The dominance relationship is defined as: For any two solutions a and b in a multi-objective minimization problem, if a is better than b in at least one objective and not worse in all others, then a dominates b. If neither a dominates b nor b dominates a, they are non-dominated with respect to each other.

Figure 9.2 illustrates the result of non-dominated sorting in a bi-objective minimization problem, showing different levels (fronts) of Pareto solutions. Solutions in the second level dominate those in the third level, while solutions in the first level dominate all others.

This non-dominated sorting method effectively selects the best solutions from the pool in multi-objective optimization problems [1], better distinguishes solution quality, ranks them, and yields a set of Pareto optimal solutions.

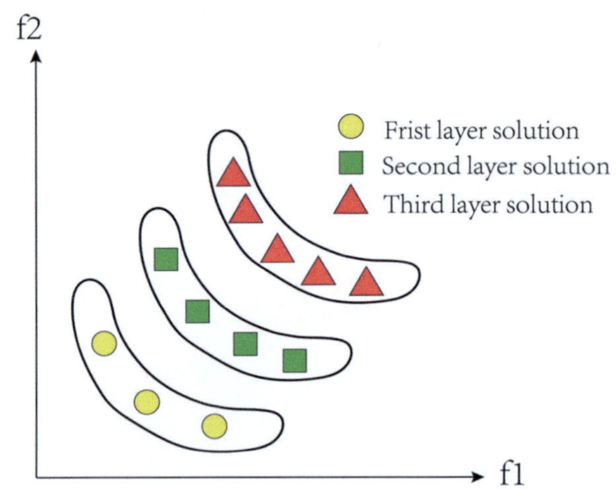

Fig. 9.2 Non-dominated sorting

9.1.2 Technique for Order of Preference by Similarity to Ideal Solution Based on Entropy Weight

After obtaining a set of Pareto solutions, it is necessary to select a single best potential point that balances global exploration and local exploitation advantages to reduce computational costs and avoid excessive calls to the true evaluation function. Solving multi-criteria decision-making (MCDM) problems effectively is crucial in many research fields. The Technique for Order of Preference by Similarity to Ideal Solution (TOPSIS) based on entropy weight [2] provides a viable approach. The core principle of entropy-weighted TOPSIS is that the ideal alternative should have the shortest geometric distance to the Positive Ideal Solution (PIS), meaning it is as close as possible to the optimal level in all objective dimensions, and the longest geometric distance to the Negative Ideal Solution (NIS), meaning it is as far as possible from the worst level. The specific steps are as follows:

(1) Construct the Normalized Decision Matrix.

For m Pareto solutions in the set, construct an initial matrix A based on the distance metric and local metric criteria mentioned in the previous subsection:

$$A = \begin{bmatrix} D_1 & \hat{f}(x_1) \\ D_2 & \hat{f}(x_2) \\ \vdots & \vdots \\ D_m & \hat{f}(x_m). \end{bmatrix} \tag{9.7}$$

In multi-criteria comprehensive decision-making, constructing a unified and reasonable evaluation index system is the prerequisite for accurate decision-making. Since different evaluation indicators often differ in scale, attributes, and value ranges, it is necessary to normalize the original indicator system to ensure comparability and meet the requirements of comprehensive decision-making. All indicators need to be converted into benefit-type indicators (where larger values are better).

For NDSM, maximizing the distance metric is a benefit-type indicator. Minimizing the local metric criterion is a cost-type indicator (where smaller values are better). To incorporate both into a unified evaluation framework, specific normalization methods are applied to transform the cost-type local metric into a benefit-type indicator. After normalization, all indicators are consistent in attribute and scale, avoiding bias caused by differences in original value magnitudes. Therefore, based on the initial decision matrix A, obtain the normalized decision matrix $B = [b_{ij}]$:

$$B = \begin{bmatrix} nD_1 & n\hat{f}(x_1) \\ nD_2 & n\hat{f}(x_2) \\ \vdots & \vdots \\ nD_m & n\hat{f}(x_m), \end{bmatrix} \tag{9.8}$$

$$nD_i = \frac{D_i - \min\{D_1, \ldots, D_m\}}{\max\{D_1, \ldots, D_m\} - \min\{D_1, \ldots, D_m\}}, \tag{9.9}$$

$$n\hat{f}(x_i) = \frac{\max\{\hat{f}(x_1), \ldots, \hat{f}(x_m)\} - \hat{f}(x_i)}{\max\{\hat{f}(x_1), \ldots, \hat{f}(x_m)\} - \min\{\hat{f}(x_1), \ldots, \hat{f}(x_m)\}}. \tag{9.10}$$

(2) Calculate Weights using the Entropy Weight Method.

The entropy weight method is used to assess the importance of each metric criterion. The entropy e_j for the j-th indicator is

$$e_j = -\frac{1}{\ln(m)} \sum_{i=1}^{m} b_{ij} \ln(b_{ij}). \tag{9.11}$$

The smaller the entropy value e_j of the j-th indicator, the larger its weight w_j. The weight w_j is calculated as

$$w_j = \frac{1 - e_j}{\sum_{j=1}^{2}(1 - e_j)}, \tag{9.12}$$

where $w_j \in [0, 1]$, $\sum_{j=1}^{2} w_j = 1$.

(3) Construct the Weighted Normalized Decision Matrix Z.

Multiply the normalized decision matrix B by the weights w_j:

$$z_{ij} = w_j b_{ij}, \quad i = 1, \ldots, m, \quad j = 1, 2. \tag{9.13}$$

(4) Determine the Positive Ideal Solution (Z^+) and Negative Ideal Solution (Z^-).

$$z_j^+ = \max\{z_{1j}, z_{2j}, \cdots, z_{mj}\}, \quad j = 1, 2, \tag{9.14}$$

$$z_j^- = \min\{z_{1j}, z_{2j}, \cdots, z_{mj}\}, \quad j = 1, 2. \tag{9.15}$$

(5) Calculate the Euclidean distance from each alternative solution to Z^+ (d_i^+) and Z^- (d_i^-):

$$d_i^+ = \sqrt{\sum_{j=1}^{2}(z_j^+ - z_{ij})^2}, \quad i = 1, 2, \cdots, m, \tag{9.16}$$

$$d_i^- = \sqrt{\sum_{j=1}^{2}(z_j^- - z_{ij})^2}, \quad i = 1, 2, \cdots, m. \tag{9.17}$$

(6) Calculate the Closeness Coefficient (T_i) for each solution:

$$T_i = \frac{d_i^-}{d_i^- + d_i^+}, \quad i = 1, 2, \cdots, m. \tag{9.18}$$

A larger T_i value indicates a larger distance metric D_i and a smaller $\hat{f}(x_i)$. In other words, within the sequential optimization algorithm framework, this point plays a key balancing role, helping the algorithm achieve an efficient balance between global exploration and local exploitation. It allows the algorithm to enhance the global prediction capability and generalization performance through exploration while reducing local errors and improving local accuracy through exploitation, thereby optimizing the overall model performance. Therefore, in selecting the new sample point for infill, the solution with the largest T_i, i.e., the highest closeness coefficient, is chosen to update the surrogate model.

9.1.3 Sequence Optimization Method Based on Non-dominated Sorting Method

This section integrates the content from Chap. 8 on multi-fidelity transfer modeling to establish a sequence optimization method based on the deep transfer multi-fidelity surrogate model: the Non-dominated Sorting Method-based Sequence Optimization Method (NDSM-SOM). The main flowchart of the method is shown in Fig. 9.3, with the specific steps outlined below:

Step 1: Generate sample points using Optimal Latin Hypercube Sampling (OLHS).

In sequential optimization, constructing a high-fidelity (HF) surrogate model does not require a large number of low-fidelity (LF) and HF samples upfront, as this would drastically increase computational cost and time. Therefore, OLHS is used to generate an appropriate number of LF and HF sample points within the design space. This method ensures that the generated sample points are as uniformly distributed as possible within the variable design space. This uniformity allows the samples to maximally cover key information in the variable space, laying a solid foundation for subsequent model construction, data analysis, and optimization decisions. It enhances the model's ability to represent and fit the characteristics of the entire design space, meeting the requirements of efficient and accurate sequential optimization.

Steps 2–3: Construct the multi-fidelity transfer learning surrogate model.

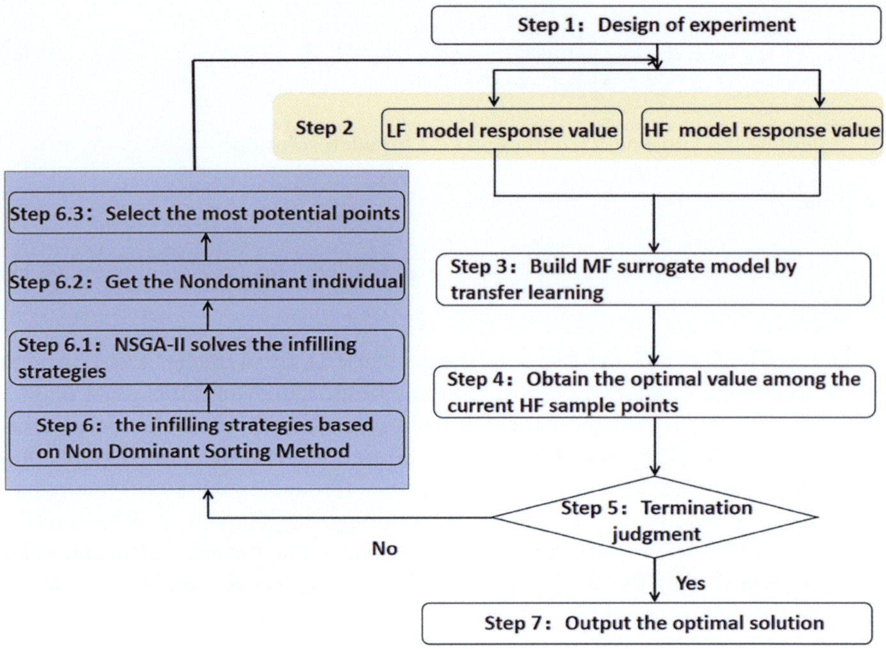

Fig. 9.3 Flowchart of the sequence optimization method based on non-dominated sorting

Based on the initially sampled LF and HF points, appropriate simulation tools are employed to obtain LF and HF response values for subsequent research. According to the TL-DTW-MFAM method described in Chap. 8, the relationship between HF and LF is expressed, and the multi-fidelity transfer learning surrogate model is constructed. This mainly includes: jointly training the feature extractor and classifier using the dynamic task weighting loss strategy based on HF and LF samples, and optimizing the network structure hyperparameters.

Step 4: Obtain the optimal value from the current HF sample points.

Select the best HF model response value from the current HF samples as the optimal solution $y_{\min}$. If the surrogate model is optimized to find the optimal solution, the model's accuracy might cause a significant deviation from the true optimal solution of the actual problem, affecting subsequent design steps.

Step 5: Evaluate termination criteria.

During sequential optimization, termination conditions can take various forms. Common ones include the maximum number of HF model evaluations and judgment based on the current optimal solution. In practical engineering applications, due to problem complexity, accurately predicting the optimal solution is difficult. To effectively control computational costs and prevent excessive resource consumption, the termination condition based on the maximum number of HF model evaluations is more widely used. In sequential optimization, fewer HF evaluations required under this condition often mean approaching the optimal solution faster and at lower cost.

This paper sets the maximum number of HF model evaluations as the termination criterion. When this criterion is met, proceed to Step 8 and output the results; otherwise, go to Step 6: Select a new sample point using the proposed NDSM infill strategy.

Step 6: Non-dominated Sorting Method-based Infill Strategy (NDSM)

The key steps in the NDSM infill strategy process within sequential optimization are non-dominated sorting and selecting the best potential point, detailed as follows:

(1) Treat global exploration and local exploitation as two conflicting objectives. The NDSM infill strategy can be expressed by Eq. 9.9, which integrates the complex relationship between these objectives and highlights the trade-offs. Solving this problem can be done using a multi-objective genetic algorithm like NSGA-II. Set appropriate algorithm parameters (crossover rate, mutation rate, population size, and optimization iteration count) to solve the infill strategy, obtaining the Pareto solutions for the global and local metrics.

(2) Select the best potential point. After obtaining the Pareto solution set, selecting the best solution from it is a crucial step to enhance the effectiveness of sequential optimization. The entropy-weighted TOPSIS method is used here for screening. This method can accurately identify the best potential point from the Pareto set. This point helps balance the relationship between global exploration and local exploitation in subsequent optimization steps, preventing premature convergence to local optima and guiding the search direction toward the global optimum, thereby improving overall optimization efficiency.

    Step 7: Output the optimal solution.

    Output the solution with the smallest target value satisfying feasibility constraints from the current HF sample set.

9.2 Example Testing and Analysis

To comprehensively verify the effectiveness of the NDSM-based sequence optimization method, this study selects two representative numerical examples for in-depth analysis of the method's characteristics in practical application. Simultaneously, comparisons are made with the EI strategy and the LCB function-based infill strategy, using the number of HF calls and the obtained optimal solution as evaluation metrics. Fewer HF calls and a smaller optimal solution indicate better infill strategy performance. This demonstrates the advantages of the NDSM-based sequence optimization method.

9.2.1 Two-Dimensional Rosenbrock Function

The EI and LCB infill strategies require a surrogate model capable of providing prediction variance (e.g., Kriging). Therefore, Kriging-MFAM (KM) is used to build the initial surrogate model for sequential optimization, denoted as KM-EI and KM-LCB. The NDSM infill strategy is also combined with Kriging-MFAM (KM-NDSM) for sequential optimization to exclude model differences. To verify algorithm robustness, each test case is repeated 10 times under the same optimization method. Figure 9.4 shows the best result from these 10 optimizations. The Rosenbrock function [3] is a classic unconstrained optimization benchmark characterized as unimodal and non-convex, posing significant optimization difficulty. Its mathematical model is

$$\begin{cases} \min f(x) = f_H(x) \\ f_H(x) = 100(x_1^2 - x_2)^2 + (x_1 - 1)^2 \\ f_L(x) = 70(x_1^2 - x_2)^2 + 0.7(x_1 - 1)^2, \end{cases} \tag{9.19}$$

where $x = [x_1, x_2]$, $f_L(x)$, and $f_H(x)$ represent the LF and HF models, with variable range $[-2.0, 2.0]^2$. Its true function is shown in Fig. 9.4a. The function has a single global minimum at $(x_1^*, x_2^*) = (1, 1)$, $f(x_1^*, x_2^*) = 0$, located within a long, narrow valley. First, sample 100 LF points and 10 HF points using OLHS. Then, build the initial variable-fidelity transfer learning surrogate model (TL-DTW-MFAM) using the dynamic task-weighted transfer learning method from Chap. 8, shown in Fig. 9.4b. Figure 9.4 shows that compared to the surrogate model built by Kriging-MFAM, TL-DTW-MFAM more accurately represents the distribution of the true function, indicating higher prediction accuracy and validating the effectiveness of the method from Chap. 8.

Set NSGA-II parameters: population size 30, generations 100, crossover probability 0.95, mutation probability 0.001, to solve the NDSM infill strategy. Set the

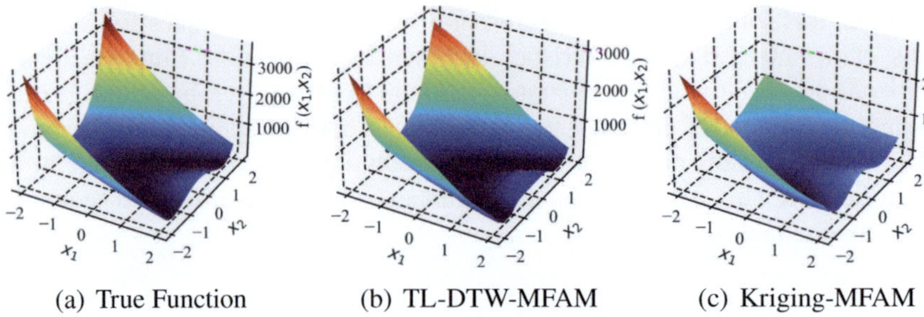

Fig. 9.4 Initial surrogate models for the Rosenbrock function

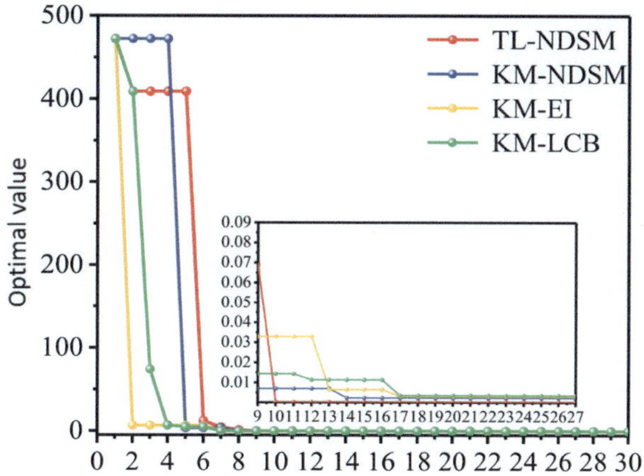

Fig. 9.5 Optimization convergence results of different methods for the Rosenbrock function

termination criterion for sequential optimization as a maximum of 30 HF evaluations. Convergence results for different sequence optimization methods are shown in Fig. 9.5.

As seen in the figure, due to the small initial sample size, the model's prediction accuracy is low. Adding new samples to update the model improves its global predictive capability. As the number of samples increases, the model's optimal solution gradually converges. If no better solution is found during optimization, the current optimal solution remains unchanged, but the surrogate model is still updated. Therefore, the convergence curve may remain flat for some time. When approaching the optimal solution, the objective function value drops significantly, showing a jump-like change. If the initial model is accurate, the algorithm converges quickly to the global optimum; if the initial model is poor, it can still find the global optimum through effective global exploration and local exploitation. TL-NDSM exhibits the fastest convergence speed, finding the optimal solution in just 10 iterations, followed by KM-NDSM, KM-LCB, and KM-EI, reaching the optimum at 12, 17, and 17 iterations, respectively. This indicates that despite differences in the initial model's prediction accuracy, NDSM, compared to traditional infill strategies EI and LCB, can more effectively balance global exploration and local exploitation during sequential optimization, guiding the process toward the global optimum with fewer sample points, thus ensuring result accuracy.

Table 9.1 shows the results of 10 random optimization runs on the Rosenbrock function. It can be seen that all infill methods perform well on the Rosenbrock function, but with some differences. Whether considering the maximum optimal solution, minimum optimal solution, or average optimal solution, the sequential optimization methods are superior to directly using a Genetic Algorithm (GA) on the HF model. For GA, population size was 20, generations 100, crossover probability 0.95, mutation probability 0.001, requiring 2000 HF evaluations. The sequence optimization methods significantly improve optimization efficiency and save computational cost. Furthermore, among the four sequence optimization methods, TL-NDSM finds the

Table 9.1 Optimization results for the Rosenbrock function

Optimization method	Max optimal obj.	Min optimal obj.	Avg optimal obj.
TL - NDSM	1.84E-4	2.06E-4	1.95E-4
KM - NDSM	6.99E-3	2.35E-3	4.67E-3
KM - EI	8.37E-3	3.11e-3	5.74E-3
KM - LCB	8.67E-3	3.52e-3	6.10E-3
GA	0.2746	5.8E-3	0.1379

optimal solution with the fastest convergence speed, thus building the most accurate surrogate model.

9.2.2 *Two-Dimensional Himmelblau Function*

The two-dimensional Himmelblau function is another classic unconstrained optimization benchmark, characterized as multimodal and non-convex, posing significant difficulty. Its mathematical model is

$$\begin{cases} \min f(x) = f_H(x) \\ f_H(x) = (x_1^2 + x_2 - 11)^2 + (x_2^2 + x_1 - 7)^2 \\ f_L(x) = f_H(0.5x_1, 0.8x_2) + x_2^3 - (x_1 + 1)^2, \end{cases} \tag{9.20}$$

where $\mathbf{x} = [x_1, x_2]$, $f_L(\mathbf{x})$, and $f_H(\mathbf{x})$ represent the LF and HF models, with variable range $[-3.0, 3.0]^2$. Figure 9.6a shows the true function surface. This function has four local minima, with the global minimum being 0. Sample 100 LF points and 10 HF points using OLHS. Build the initial variable-fidelity transfer learning surrogate model (TL-DTW-MFAM) using the dynamic task-weighted transfer learning method from Chap. 8, shown in Fig. 9.6b. Figure 9.6 shows that both TL-DTW-MFAM and Kriging-MFAM surrogate models fit the true function distribution well, indicating similar high prediction accuracy for the initial models.

For solving the NDSM infill strategy, set NSGA-II parameters: population size 30, generations 100, crossover probability 0.95, and mutation probability 0.001. Set the termination criterion as a maximum of 30 HF evaluations. Convergence results for different sequence optimization methods are shown in Fig. 9.7. TL-NDSM exhibits the fastest convergence speed, finding the optimal solution in 13 iterations.

KM-NDSM, KM-LCB, and KM-EI find the optimum at 17, 25, and 26 iterations, respectively. Table 9.2 shows the results of 10 random optimization runs on the Himmelblau function. TL-NDSM outperforms other methods in terms of maximum, minimum, and average optimal objective values. This again demonstrates that TL-NDSM achieves the best optimization performance while ensuring accuracy and exhibits the fastest convergence speed.

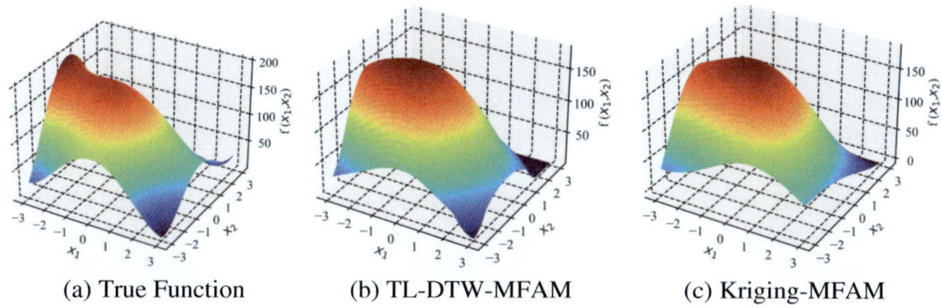

Fig. 9.6 Himmelblau function and initial surrogate models

Fig. 9.7 Optimization convergence results of different methods for the Himmelblau function

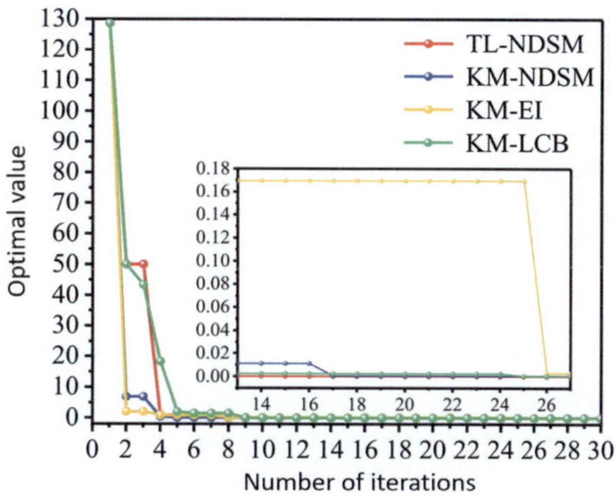

Table 9.2 Optimization results for the Himmelblau function

Optimization method	Max optimal obj.	Min optimal obj.	Avg optimal obj.
TL - NDSM	7.57E-5	6.61E-5	7.09E-5
KM - NDSM	7.99E-4	4.94E-4	6.47E-4
KM - EI	3.62E-3	3.03E-3	3.33E-3
KM - LCB	1.89E-3	4.74E-4	1.18E-3
GA	1.138	0.665	0.901

Through the comprehensive validation with two numerical examples, the results demonstrate that the proposed TL-NDSM sequence optimization method outperforms traditional sequence optimization methods based on EI and LCB infill strategies. This method effectively balances global exploration and local exploitation during optimization, reduces the number of new sample points required, effectively guides the optimization process toward the global optimum, ensures result accuracy, and significantly saves computational cost.

9.3 Summary

To address the limitations of traditional Gaussian-process-based sequence optimization methods in balancing global exploration and local exploitation, this chapter proposed a novel sequence optimization method based on the non-dominated sorting method and the multi-fidelity transfer learning surrogate model. In this method, global exploration is achieved through a distance metric, and local exploitation is achieved using the current best value metric. These two are treated as conflicting objectives, and their Pareto solutions are solved using NSGA-II. Subsequently, the best potential point is selected using entropy-weighted TOPSIS to update the surrogate model. Validation through two numerical examples shows that the proposed method effectively ensures model accuracy while significantly improving optimization efficiency.

References

1. Deb K, Pratap A, Agarwal S et al (2002) A fast and elitist multiobjective genetic algorithm: NSGA-II. In: IEEE Trans Evol Comput 6(2):182–197. https://doi.org/10.1109/4235.996017
2. Tzeng GH, Huang JJ (2011) Multiple attribute decision making: methods and applications. CRC Press
3. Rosenbrock HH (1960) An automatic method for finding the greatest or least value of a function. Comput J 3:175–184. https://doi.org/10.1093/comjnl/3.3.175

Chapter 10
Conclusion Remarks

10.1 Conclusions

This monograph provides a comprehensive introduction to the use of variable-fidelity surrogate models in engineering design and optimization.

Chap. 1 sets the foundation by discussing the challenges in modern product design optimization, particularly with high-fidelity computational simulations such as FEM and CFD, which offer accuracy but at high computational costs. It introduces surrogate-based design optimization as a solution, explaining surrogate models including RSM, RBF, Kriging, ANN, MARS, and SVR, along with their applications in multidisciplinary, robust, and global optimization. The chapter then presents variable-fidelity surrogate models, defining fidelity levels, their sources, for instance, mesh resolution or physical assumptions, and the inverse relationship between fidelity and computational efficiency. It outlines the overall framework, covering DOE methods such as Latin Hypercube Design (LHD), evaluation, model construction, optimization strategies including offline and online SBDO, algorithms, and optimal solution evaluation. Finally, the chapter reviews literature on experimental designs for variable-fidelity surrogates, such as nested LHD and adaptive sampling; surrogate modeling approaches, including linear and nonlinear scaling as well as deep learning integrations like variable-fidelity deep neural networks and variable-fidelity physics-informed neural networks; and optimization strategies, such as trust region-approximation model management optimization and efficient global optimization with infill criteria including EI and proportional expected improvement.

Chap. 2 introduces key technologies in variable-fidelity surrogate modeling. It begins with the fundamental concepts and workflow of surrogate models, followed by detailed explanations of common models, including RSM, MARS, RBF, Kriging, and SVR. These models are discussed in terms of their principles, characteristics, and applicable scenarios. It then covers variable-fidelity modeling techniques, including additive and multiplicative scaling methods, the construction of global scaling functions using Kriging, and the TR-AMMO optimization strategy. For high-dimensional

J. Yi et al., *Variable-Fidelity Surrogate*, Engineering Applications of Computational Methods 25, https://doi.org/10.1007/978-981-95-5527-7_10

problems, High-Dimensional Model Representation (HDMR) is introduced. Finally, the chapter discusses model validation methods, evaluation metrics (e.g., RRMSE, RAAE, and RMAE), and experimental design strategies, including classical space-filling designs (like LHD) and adaptive sequential sampling, emphasizing three key principles in variable-fidelity design: nesting, economy, and space-filling.

Both Chaps. 3 and 4 discuss the experimental design for variable-fidelity surrogate modeling. Chapter 3 introduces an experimental design method for variable-fidelity surrogate modeling—the translation propagation-based nested Latin hypercube design. The approach begins with a simple nested seed design, which is systematically translated and propagated across the design space to generate large-scale nested sample sets. Its main advantages lie in its simplicity, computational efficiency, and its ability to naturally satisfy the principles of nested structure and economy while maintaining a degree of space-filling. To accommodate arbitrary sample sizes, a point-removal strategy is proposed, where excess points are eliminated based on their distance from the design center to preserve nested relationships. Through numerical benchmarks and an engineering case study, the effectiveness of the method is validated. Results show that surrogate models constructed using this nested design method consistently outperform those based on non-nested designs in terms of prediction accuracy, confirming the method's capability to generate rational sample distributions and support high-quality variable-fidelity model construction.

Chapter 4 proposes a nested maximin design method, named SLE-NGHS, for constructing variable-fidelity surrogate models. The method first generates low-fidelity sampling points using the Successive Local Enumeration (SLE) method, which ensures good space-filling properties, and then selects a subset of these points as high-fidelity samples using an improved NGHS algorithm to maximize the minimum inter-point distance. Comparative numerical experiments against existing methods demonstrate the superiority of SLE-NGHS in terms of space-filling performance, computational cost, and prediction accuracy. Furthermore, the method is validated in an engineering case study on the optimal design of a natural gas pressure vessel, confirming its effectiveness and practical applicability.

Chapters 5–7 focus on the variable-fidelity surrogate modeling methodologies. Chapter 5 proposes a variable-fidelity surrogate modeling method based on scale functions, termed "Difference Mapping Framework with Least Squares Support Vector Regression (DMF-LSSVR)." The process begins by designing experiments on both high-fidelity and low-fidelity models to obtain sample data. Through gap mapping, the low-fidelity model is stretched toward the high-fidelity one, and LSSVR is used to learn the discrepancy between the two. The resulting surrogate model effectively integrates the strengths of both fidelity levels, especially under limited data conditions. The method is validated through three engineering case studies—the deep-hole model, vortex generator design for inverted wings, and heat exchanger design—and compared against single-fidelity and linearly corrected multi-fidelity models. Results show that DMF-LSSVR consistently achieves higher prediction accuracy, particularly under nested experimental designs.

Chapter 6 introduces the Variable-Fidelity Physics-Informed Neural Network (VF-PINN) framework, which enhances standard PINNs by integrating low-fidelity analytical or coarse-grid models with sparse high-fidelity data to solve partial differential equations (PDEs) with improved accuracy and reduced computational cost. The chapter details the multi-network architecture, including separate low- and high-fidelity networks fused through a residual correction mechanism, and formulates a composite loss function that balances PDE residuals, boundary conditions, and multi-fidelity data alignment. Through detailed case studies on the Euler-Bernoulli beam with variable point loads and the two-dimensional Poisson equation with variable source intensity, the chapter evaluates VF-PINN's performance using metrics such as MAPE, RMSE, and R^2, demonstrating its superiority over standard PINNs and standalone fidelity models in terms of error reduction, generalization to unseen parameters, and efficiency. The chapter concludes by affirming VF-PINN's versatility for engineering applications, providing a robust tool for variable-fidelity surrogate modeling in design optimization.

Chapter 7 introduces a Multi-Fidelity Transfer Learning Model based on Dynamic Task-Weighted Loss (HDR-DTW) to address the challenge of limited high-fidelity data in surrogate modeling. This method integrates deep transfer learning with powerful nonlinear neural network architectures to jointly extract shared features and fit tasks for both low- and high-fidelity models. A harmonized network structure consisting of a feature extractor and two task-specific regressors is proposed. The novel dynamic task-weighting strategy adjusts the contribution of source and target tasks throughout training to overcome issues such as conflicting gradients, varying convergence rates, and task dominance. Experiments on a one-dimensional function and ten multidimensional benchmark problems demonstrate that HDR-DTW achieves superior predictive accuracy and robustness compared to existing MF surrogate models, especially in high-dimensional, data-scarce scenarios. The method's practical value is further validated through an engineering case study on optimizing a permanent magnet gear, where HDR-DTW significantly reduces computational cost while maintaining high accuracy.

Chapters 8 and 9 discuss the variable-fidelity surrogate-assisted optimization approaches. Chapter 8 introduces an online Variable-fidelity Surrogate-assisted Harmony Search algorithm (VFS-HS) designed to tackle computationally expensive engineering design optimization problems. By integrating high- and low-fidelity models into surrogate functions and employing a dynamic sampling strategy with multi-level screening, the algorithm effectively balances global exploration and local exploitation. Experimental results on benchmark functions and a real-world engineering case demonstrate that VFS-HS achieves superior optimization performance, robust convergence, and strong stability, outperforming several state-of-the-art single- and multi-fidelity surrogate-assisted optimization methods. These findings highlight its potential for solving complex, high-cost optimization tasks.

Chapter 9 presents a novel optimization strategy based on the non-dominated sorting method and a multi-fidelity transfer learning surrogate model, aimed at addressing computationally expensive design optimization problems. The method treats global exploration and local exploitation as conflicting objectives, formulates a bi-objective infill criterion using distance and minimum surrogate prediction, and employs NSGA-II to obtain infilling points. The best sample point is selected via entropy-weighted TOPSIS to update the surrogate. Experimental validation on two classical benchmark functions demonstrates that this method achieves faster convergence, better accuracy, and lower computational cost compared to traditional sequential strategies such as EI and LCB, highlighting its robustness and applicability in complex engineering optimization tasks.

10.2 Future Works

While this monograph has laid a solid foundation for variable-fidelity surrogate modeling and its applications in engineering design and optimization, several promising avenues remain for future research. These directions aim to address current limitations and leverage emerging technologies to enhance the robustness, scalability, and applicability of variable-fidelity methods.

One critical area is the integration of Uncertainty Quantification (UQ) into variable-fidelity surrogate frameworks. Current approaches often handle noise and discrepancies between fidelity levels, but comprehensive UQ techniques such as Bayesian inference or ensemble methods could be further developed to propagate uncertainties across fidelities more accurately. This would enable better risk assessment in optimization processes, particularly in safety-critical applications like aerospace and structural engineering, where quantifying prediction confidence is essential for decision-making.

Building on the VF-PINN framework introduced in Chap. 6, future research should prioritize the deeper integration of physical information into variable-fidelity surrogate modeling paradigms. The incorporation of physics-informed constraints is pivotal, as it not only enforces strict adherence to fundamental governing equations such as PDEs in fluid dynamics or structural mechanics but also bolsters model robustness and interpretability, particularly in scenarios involving sparse or noisy data. This physics-guided approach effectively addresses the shortcomings of purely data-driven methods, which frequently falter in extrapolating beyond their training regimes, by seamlessly embedding domain-specific knowledge directly into the neural network architecture. Notably, VF-PINN demonstrates substantial potential for tackling approximation modeling challenges in high-dimensional spaces, enabling the construction of highly accurate models with significantly fewer sample data points, thereby enhancing efficiency in computationally intensive engineering applications.

Another exciting direction is the extension of variable-fidelity surrogates to digital twin technologies. Traditional applications have emphasized static performance predictions, but future developments should shift toward dynamic modeling over the entire product lifecycle. This includes real-time updating of surrogates with sensor data from operational phases, enabling predictive maintenance, fault detection, and adaptive optimization. Integrating variable-fidelity models with digital twins could revolutionize fields like manufacturing and energy systems, where continuous monitoring and simulation fusion allow for lifecycle-aware decision-making.

Addressing high-dimensional design spaces remains a challenge for variable-fidelity surrogate modeling. Future research could explore dimensionality reduction techniques, such as active subspaces or variational autoencoders, combined with multi-fidelity data fusion to handle problems with hundreds or thousands of variables. This would expand the applicability to domains like materials science and genomics, where high-dimensionality is inherent, and efficient surrogates could significantly cut computational costs.

Lastly, a promising extension involves continuous fidelity levels, where fidelity can be finely tuned within a single data source through adjustable parameters. For instance, in FEA or CFD simulations, mesh sizes can be continuously varied to control fidelity, allowing for seamless integration across a spectrum of accuracy levels. Data fusion methods and neural networks, such as transfer learning and generative models, can be employed to aggregate information from these varying fidelity points. Investigating variable-fidelity surrogate modeling under such continuous fidelity scenarios could enable more precise cost control while maintaining model accuracy, optimizing resource allocation in computationally intensive designs like structural and fluid dynamics applications.

MIX
Papier aus verantwortungsvollen Quellen
Paper from responsible sources
FSC® C105338

If you have any concerns about our products,
you can contact us on
ProductSafety@springernature.com

In case Publisher is established outside the EU,
the EU authorized representative is:
Springer Nature Customer Service Center GmbH
Europaplatz 3, 69115 Heidelberg, Germany

Printed by Libri Plureos GmbH
in Hamburg, Germany